Glimpses of Space Exploration

Glimpses of Space Exploration

Sadasiva Biswal

PRINTWORLD
Publishers of Indian Traditions

Cataloging in Publication Data — DK
[Courtesy: D.K. Agencies (P) Ltd. <docinfo@dkagencies.com>]

Biswal, Sadasiva, 1945- **author.**
Glimpses of space exploration / Sadasiva Biswal.
pages cm
Includes bibliographical references and index.
ISBN 9788124613085 (HB)

1. Outer space – Exploration. 2. Space vehicles.
3. Astronautics. I. Title.

LCC QB500.262.B57 2025 | DDC 523.2 23

ISBN: 978-81-246-1308-5

First published in India in 2025

Printed and published by:
D.K. Printworld (P) Ltd.
Regd. Office : "Vedaśrī", F-395, Sudarshan Park
ESI Hospital Metro Station, New Delhi - 110015
Phones : (011) 4172 2626, 98713 94815
e-mail : indology@dkprintworld.com
Web : www.dkprintworld.com

To
all my revered teachers
who have shown me the path of learning

Prologue

SPACE exploration is the use of astronomy and space technology to explore outer space. While the exploration of space is carried out mainly by astronomers with telescopes, its physical exploration though is conducted both by uncrewed robotic space probes and human spaceflight.

Space exploration is investigation, by means of crewed and uncrewed spacecraft, of the reaches of the universe beyond the earth's atmosphere and the use of the information so gained to increase knowledge of the cosmos and benefit humanity. Space exploration, like its classical form astronomy, is one of the main sources for space science.

Space exploration deals with space science and human presence in space, and that is carried out by spaceflights and astronauts. Spaceflight is the use of space technology to achieve the flight of spacecraft into and through outer space. Spaceflight is used for space exploration and also for commercial activities like space tourism and satellite telecommunications. Additional non-commercial uses of spaceflight include space observatories, reconnaissance satellites, and other earth observation satellites.

Exploration of space has been one of the fundamental human instincts. For centuries, man has marvelled at the night sky and has been devising several means to go beyond the earth and reach the universe; it was only in the past two centuries, exciting developments have been taken place in the areas of spaceflight and space exploration. As a result, man could finally be able to go around the earth, land on the moon and send spacecraft to other planets in the solar system and beyond. We have also acquired capabilities to have a clear view of the universe through telescopes orbiting around the earth like the Hubble and Chandra observatories.

Though space exploration began with man's quest for knowledge, the advanced technologies paved the way for the application of space missions and explorations to the development of the society. Today, sapce-based telecommunication, television broadcasting, meteorology, resources monitoring, etc. have become essential in everyday life. Ostensibly, space exploration has played a key role in our ultra-modern society providing luxury to human life. The fascinating but complex technology of space can be understood only when it is conveyed in a lucid and interesting manner. Everyday, some amazing feats are happening in space science technology and its applications.

In the present treatise, *Glimpses of Space Exploration*, attempts have been made to enunciate the complexity of space technology in the simplest forms for the easy understanding of common man interested in space. Different aspects of space missions and the technology underlying there have been spread over thirty-six chapters, beginning with a Prelude and ending with an Epilogue. The contributions of women in space travel and exploration have also been discussed. Both advantages and disadvantages of space missions form a part of our discourse. A select Bibliography is given for further reading.

Most of the topics presented here have been published earlier in magazines/journals like *Science Horizon*, *Emerging Science,* and *Current Development Reporter,* and received readers' applaud. I am highly indebted to those editors and the authorities for inspiring me to present all those articles collectively in the form of a book.

I have also drawn some material from the books of renowned authors and publishers. I owe my indebtedness to all of them. Aknowledgement is also due to my revered teachers and loving pupils for their inestimable help and prolonged initiations. It is a pleaure on my part to recognize their assistance.

I am ever indebted to my parents for their eternal blessings on every walk of my life. Their sweet compassion, love, and affection always provide me enough strength to walk through all odds. All of my achievements go to them.

I feel incompetent in acknowledging the role of my wife late Dr Bidyut Lata Ray, who had constantly encouraged me and supported me in every step of my life. I am ever indebted to her.

Lastly, I render my sincere obligations to D.K. Printworld for bringing out the book with all caution and care. Any suggestion regarding the improvement of the book is most welcome.

Knowledge dispels ignorance.

Sadasiva Biswal

Contents

1

Prelude

Space exploration is the use of astronomy and space technology to study and traverse the outer space. While the exploration of space is carried out mainly by astronomers with telescopes, its physical exploration is conducted both by uncrewed robotic space probes and human spaceflight.

Space exploration is an investigation, by means of crewed and uncrewed spacecraft, of the reaches of the universe beyond earth's atmosphere and the use of the information so gained to increase the knowledge of the cosmos and benefit humanity. Space exploration, like its classical form astronomy, is one of the main sources for space science.

Humans have always looked at the sky and wondered about the nature of the objects seen in it. With the development of rockets and the advances in electronics and other technologies in the twentieth century, it became possible to send machines and animals, and then people above earth's atmosphere into the outer space. Well before technology made these achievements possible, however, space exploration had already captured the minds of many people, not only of aircraft pilots and scientists but also of writers and artists. The stronghold that space travel has always had on the imagination may well explain why professional astronauts and laypeople alike consent at their great peril, in the words of

Tom Wolfe in *The Right Stuff* (1979), to sit "on top of an enormous Roman Candle, such as a Redstone, Atlas, Titan, or Saturn rocket, and wait for someone to light the fuse". It perhaps also explains why space exploration has been a common and enduring theme for literature and art. As centuries of speculative fiction in books and more recently in films makes clear, "one small step for a man, one giant leap for mankind" was taken by the human spirit many times and in several ways before Neil Armstrong stamped humankind's first footprint on the moon.

While the observation of objects in space, known as astronomy, predates reliable recorded history, it was the development of large and relatively efficient rockets during the mid-twentieth century that allowed the physical space exploration to become a reality. The world's first large-scale exprimental rocket programme was Opel-RAK under the leadership of Fritz von Opel and Max Valier during the late 1920s leading to the first crewed rocket cars and rocket planes, which paved the way for the Nazi era V2 programme and the US and Soviet activities from 1950 onwards. The Opel-RAK programme and the spectacular public demonstrations of ground and air vehicles drew large crowds, that caused global public experiment as the so-called "Rocket Rumble" and had a large long-lasting impact on later spaceflight pioneers like Wernher von Braun. Common rationales for exploring space include advancing scientific research, national prestige, uniting different nations, ensuring the future survival of humanity, and developing military and strategic advantages against other countries.

Achieving spaceflight enabled humans to explore the solar system and the rest of the universe, to understand many objects and phenomena that are better observed from a space perspective, and to use for human benefit the resources and attributes of the space environment. All of these activities — discovery, scientific understanding, and the application of that understanding to serve human purposes — are the elements of space exploration.

Although the possibility of exploring space has long excited people in many walks of life, for most of the latter twentieth century and into the early twenty-first century, only national governments could afford the high costs of launching people and machines into space. This reality meant that space exploration had to serve broad interests, and it indeed has done so in a variety of ways. Government space programmes have increased knowledge, served as indicators of national security and military strength, and provided significant benefits to the general public. In areas where the private sector could profit from activities in space, most notably the use of satellites as telecommunication relays, commercial space activity has flourished without government funding. In the early twenty-first century, entrepreneurs believed that there were several other areas of commercial potential in space, most notably privately-funded space travel.

After Second World War, governments assumed a leading role in the support of research that increased fundamental knowledge about nature, a role that earlier had been played by universities, private foundations, and other non-governmental supporters. This change came for two reasons. First, the need for complex equipment to carry out many scientific experiments and the large teams of researchers to use that equipment led to high costs that only governments could afford. Second, governments were willing to take on this responsibility because of the belief that fundamental research would produce new knowledge essential to the health and security domains, and would enhance the quality life of their citizens. Thus, when scientists sought government support for early space experiments, it was forthcoming. Since the start of space efforts in the United States, the Soviet Union, and in Europe, governments have given high priority to the support of science done in and from space. From modest beginnings, space science has expanded under government support to include multibillion-dollar exploratory missions in the solar system. Examples of such efforts include the development of the curiosity Mars rover, the Cassini–

Huygens mission to Saturn and its moons, and the development of major space-based astronomical observatories such as the Hubble Space Telescope.

In addition to providing security benefits, satellites offered military forces the potential for improved communications, weather observation, navigation, timing, and position location. This led to significant government funding for military space programmes in the United States and Soviet Union. Although the advantages and disadvantages of stationing force-delivery weapons in space have been debated, as of the early twenty-first century, such weapons had not been deployed nor had space-based anti-satellite systems, i.e. systems that can attack or interfere with orbiting satellites. The stationing of weapons of mass destruction in orbit or on celestial bodies is prohibited by international law.

Nations realized early on that the ability to observe the earth from space could provide significant benefits to the general public apart from security and military uses. The first application to be pursued was the development of satellites for assisting in weather forecasting. A second application involved remote observation of land and sea surfaces to gather imagery and other data of value in crop forecasting, resource management, environmental monitoring, and other applications. The United States, the Soviet Union, Europe, and China also developed their own satellite-based Global Positioning Systems (GPS), originally for military purposes, that could pinpoint a user's exact location, help in navigating from one point to another, and provide very precise time signals. These satellites quickly found numerous civilian uses in such areas as personal navigation, surveying and cartography geology, air-traffic control, and the operation of information-transfer networks. They illustrate a reality that has remained constant for a half century — as space capabilities are developed, they often can be used for both military and civilian purposes.

Another space application that began under government sponsorship but quickly moved into the private sector is the relay of voice, video, and data via orbiting satellites. Satellite telecommunication has developed into a multibillion-dollar business and is the one clearly successful area of commercial space activity. A related, but economically much smaller, commercial space business is the provision of launches for private and government satellites. In 2004, a privately financed venture sent a piloted spacecraft, SpaceShipOne, to the lower edge of space for three brief suborbital flights. Although it was technically a much less challenging achievement than carrying humans into orbit, its success was seen as an important step towards opening up space to commercial travel and eventually to tourism. More than fifteen years after SpaceShipOne reached space, several firms began to carry out such suborbital flights. Companies have arisen that also use satellite imagery to provide data for business about economic trends. Suggestions have been made that in the future other areas of space activity, including using resources found on the moon and near-earth asteroids and the capture of solar energy to provide electric power on earth, could become successful businesses.

2

The Dream and the Quest for Great Aviation

Aviation is the act of flying in the air. In the ancient times, it was a dream to fly in the air or to cross a river, a lake, and seas by swimming. Early man was perfectly safe on land. It took him quite some time to learn swimming. By flexing his limbs, he stayed above water and steered himself across. By doing this, he became tired.

In course of time, man learned from nature to move through water. He watched logs of wood floating in water, and from this he got an idea. He sat atop a log. It stayed afloat. He rode a log and forded a river in spate. It became easy. It did not tire him out. With this experience, the man advanced to prepare canoes from the logs. Sitting on the canoes and working the oars, he was able to cross waterbodies. Gradually, he made boats of wood or reeds and sailed the high seas. Then came ships that sailed by wind power. Soon motorboats and steamships were designed and man crossed waterbodies easily. Water did not limit his movements any longer.

Observing the flying birds, man imagined to have wings to fly in air. He studied the birds' motion and kept a record of it. Many facts came to light. Such observations reveal that every bird has wings. It flaps the wings and flies up and down. The bird kite flaps its wings, soars to great heights, spreads its wings, stops flapping them and glides gracefully.

Man's studies gave wings to his imagination. He shaped angels after the human form and lent them wings. For he believed that wings held magical powers. Those with wings could fly.

Human being's desire for flying in the air continued for a long period. Many innovators tried and failed. Some also lost their lives during the trial. In 1130 CE, the Benedictine Monk Eilmer of Malmesbury, Wiltshire, attached wings and took off from a mount. He crashed and broke his legs. In 1530, Giovanni Battista Danti tried to fly with wings, but he lost his life.

Leonardo da Vinci , the famous artist, was an innovator too. He designed a birdlike flying machine. He tried the airscrew machine, which is the forerunner of the helicopter. He tested the design of a parachute. Heavy objects drifted slowly to the earth when held by strings, above which spread huge canopies made of cloth.

In 1610, Francesco Lana de Terzi guessed that objects needed immense buoyancy to resist gravity. He designed a flying machine with a ship's hull, a mast, and four globes out of which air had been pumped out to make them lighter. It did not fly. But he was on the right track. In 1853, Sir George Cayley built a glider. Its wing surface was 200 sq. ft. It had three pairs of wings and a tailpiece. The wings curved more on top than on the bottom. This design was backed by Bernoulli's principle, i.e. an object with a flat base and a curved top reduces air pressure on top and that results in upward thrust. Cayley mounted the glider on wheels. It had room for a man. Cayley's coachman got on to the glider. The glider was pulled fast across a vast open ground. It gained the speed needed to take off.

William Henson called Cayley "The Father of Aviation". He improved the design of Cayley by adding two pusher propellers and managed a short flight.

Then Alphonse Pénaud produced working models, powered by twisted rubber motors. Such models had only one flaw. Those did not remain stable while taking turns. For that, moving wings were required.

Otto Lilienthal shaped gliders whose wings could be bent and manipulated to maitain stability. He made over 2,000 flights. His flights were photographed. All went well till 9 August 1896. On that day Otto took off in the glider. He soared into space. Suddenly the air current weakened and the glider lost movement. Adjustments of his weight by shifting could not be fruitful. Otto met with an accident and suffered grievous injuries and died that night. His last words were: "Sacrifices must be made".

In 1896, Samuel Langley made two models. The models flew well under normal conditions. In 1903, Langley built a full-size machine. After testing it twice, the engine failed. Langley came very close to success. But success eluded him.

After a few days of Langley's unsuccessful attempts, two young men, Wilbur Wright and Orville Wright, sons of a priest, reached the winning post.

In 1783, men first soared into space, riding a hot-air balloon. It was the result of pioneering work by Joseph-Michel Montgolfier and his brother Jacques Étienne of Annonay, France. They realized that a balloon, filled with hot air, gained the required upward thrust for flight because hot air is lighter than the air around us.

Whether an object heavier than air can fly defying gravity was a question remained unanswered for over 120 years. During these years, many people in Europe and the United States experimented with kites and gliders. These experiments proved that objects heavier than air could fly. But nobody knew how to engineer a flying machine that would remain airborne.

Many innovators tried and failed. Failures, it is said, are stepping stones to success. Finally, two young men, Wilbur Wright and his brother Orville Wright, became successful in designing a machine to fly. That machine was heavier than air. Though the machine was crude and unsophisticated, it flew. This event marked the beginning of the great aviation story.

Daring men and women explored the frontiers of the aircraft's range and speed by designing shapes and engines based on latest technological advances. The explorers continued their effort to take the aircraft to new records in speed, safety, style and elegance.

Aircrafts cannot take off or land almost anywhere. They need long stretches of runways. The quest for an alternate machine that would take off and land vertically led to the invention of the helicopter. Innovators are still working for the development of the style, design and performance of the aircrafts.

3

The Shape of the Earth

People of several cultures have studied the shape of the planet earth down the ages. Ancient astronomical studies reveal calculations of the earth's diameter and circumference. In the seventh century, Brahmagupta gave an almost accurate measure of the cirumference of the earth. Greek scholars imagined the earth as a sphere as perfect as the sun and the moon. In the sixth century BCE, Pythagoras' idea on the shape of the earth was popular. In the third century BCE, the finest achievement of ancient Greece came to light, when Eratosthenes (276–194 BCE) of Egypt determined the cricumference of the earth with remarkable accuracy. He observed that the sun shone right down a well in Syenne (Aswan) in mid-summer while at Alexandria, which was almost due north, it cast a shadow marking an angle of 7.2° to the vertical or one-fiftieth of a circle. He concluded that the earth's circumference should be fifty times the distance from Aswan to Alexandria. He calculated the distance between the two towns on the basis of a camel's average speed. As a result, he came within just 1 per cent of the correct value of the earth's circumference, viz. 39,852 km as against the present-day measurement of about 40,075 km.

Āryabhaṭṭa, born in 476 CE, was an intellectual rebel of India. At the age of 23, he became famous in Kusumapura, which is now

Patna. He boldly described the earth as spherical, circular in all directions.

About 1,000 years before Copernicus, Āryabhaṭṭa declared that the earth rotated on its axis and went around the sun. Bhāskara, a follower of Āryabhaṭṭa, composed his commentary of *Āryabhaṭīyam* in 629 CE. He obtained the mean celestial longitudes of planets by a simple and short-cut technique. His works have been well known among astronomers for centuries, especially in Kerala.

Varāhamihira (b.587 CE) and Brahmagupta (628 CE) did not accept the idea of Āryabhaṭṭa. Bhāskarācārya, an astronomer of the sixth century CE, calculated the time taken by the earth to orbit the sun to nine decimal places, almost accurate by modern measurement.

Ideas on the earth's orbit changed slowly in Europe. The Polish astronomer, Nicolaus Copernicus (1473–1543) observed that the earth rotated on its axis and went in an annual orbit around the sun. He, however, describrd the orbits of the planets as circular. A Danish astronomer, Tycho Brahe (1546–1601) pointed out that planets did not move in perfect circles. His assistant, Johannes Kepler (1571–1630) studied the ideas and stated that the orbits of planets were elliptical. He also held that as the altitude increases, a planet takes longer to complete an orbit and the planetary speed also varies with the distance from the sun.

Above discussions lead us to arrive at the following concepts about the shape of planet earth.

Flat Earth

The concept of "flat earth" was deeply rooted in the minds of the Babylonians. To them the earth was of a rounded shape having flat area. It was considered to be higher in the centre than at the edges, and to be floating on water. This could suggest that the earth was like a disc to the Babylonians (Ronan 1964: 232).

The Egyptians also have a long history. Yet, in the field of science and more particularly in the science of astronomy, they contributed nothing (Murthy 1997: 247). To the Egyptians, heavens were shaped like a dome and the earth was flat and oblong in shape. So far their conception of things appears to differ little from that of the Babylonians.

Disc-like Earth

The Greeks believed in a disc-like earth. The early Greeks were influenced by some of the Egyptian ideas. In brief, the earth appeared to them as a flat disc. The earth was fixed in the middle of the whole sphere of creation.

Spherical Earth

Anaximander (Murthy 1997: 248) has accepted the above views but with a little change. Ronan states that Pythagoras accepted the idea of a spherical universe, but rejected the belief that the earth was a flat disc. Pythagoras, instead, claimed that the earth itself was a sphere. It may be noted that Pythagoras lived about fifth-sixth century BCE for Ronan observes that "it also seems evident that he (Pythagoras) has not born before 570 BCE and he died around 497 BCE".

Plato, who lived between the first quarter of the fifth century BCE and the middle of the fourth century BCE, accepted the Pythagorean idea of a spherical earth, stationary in the centre of a spherical universe. Aristotle also thought that the universe was spherical with the earth stationary at its centre. Aristotle was the first to suggest that the earth is round for during an eclipse of the moon, the shadow of the earth is curved.

Eratosthenes, in the third century BCE, calculated the circumference of the earth, which is close to the modern value. Ptolemy places himself fairly and squarely on the side of Aristotle and accepts a spherical earth immovable in the centre of a spherical universe (Ronan 1964: 232). Considering the *pañca-mahābhūta* theory of the evolution of the universe, Varāhamihira in his

Pañcasiddhāntika (Thibaut and Dwivedi 1968: 105) described the earth:

> The spherical earth, which is constituted of the five elements, stands poised in the region of space as if it is an iron ball held in position in a cage of magnet.

The same opinion is also expressed by Lalla (Subbarayappa and Sarma 1985: 338). According to him, this sphere of earth, made of *ākāśa*, air, fire, water, and clay, and thus having all the properties of the five elements, surrounded by the orbits (of the moon, etc.) and extending up to the sphere of stars, remains on the centre of space.

Just as a ball of iron , when suspended amidst pieces of magnets, remains suspended in space, like that this globe of earth remains in space unsupported, while itself remaining the abode of all.

The Concept of Global Earth

The earth has been described as a globe in many astronomical texts. The term *bhūgola* is first found in the *Pañcasiddhāntikā* by Varāhamihira (sixth century CE). It is probable that the word featured in the *Sūrya-Siddhānta*, which was there prior to Varāhamihira. That would be the period from the *Vedāṅga Jyotiṣa* of Lagadha to the time of Varāhamihira. With its base of vast information from the Vedic, Vedāntic, and Brāhmaṇa literatures, the concept of the global earth has been buried and it is difficult to cull out the exact reference at present. Since the Ṛgvedic age ended with the battle of the ten kings in *c.*3730 BCE, the concept of the global earth must be as old as this period in the Hindu astronomical texts (Rajaram 1993).

Conclusion

The change in the concept of the disc-like earth to spherical earth took place in Europe about fifth/sixth century BCE during the period of Pythagoras. Even then, they did not know how the spherical earth could exist in space unsupported until Isaac Newton explained it by his Theory of Gravitation in 1686 CE.

The concept that the earth is in the shape of a globe and is just suspended in space like a magnet was something novel even at the late nineteenth century CE. The notion of global earth has been supported only in this century by the advances made in astronomy, space science, geophysics, etc.

Modern calculations give a little less than 40,000 km as the circumference of the earth. The diameter of the earth is about 13,155 km or about 8,220 miles. To someone, the radius of the earth is nearly 6.37×10^6 ms or 6,370 km (Weber et al. 1977: 173).

4

The Earth's Shape from Orbital Background of Satellites

Prelude

Isaac Newton was the first scholar to suggest an artificial earth satellite. In his *Principia* (1687), he explained how an object could go into orbit around the earth, if it maintained adequate speed to counter the earth's gravity. He pointed out that his Law of Universal Gravitation pervades nature and governs the motions of all objects.

In 1870, Edward Hale (1822–1909) a clergyman in Boston, USA, suggested the use of an artificial satellite to assist in navigation. It was then considered science fiction.

On 4 October 1957, the then Soviet Union placed an artificial satellite Sputnik-1, into an orbit around the earth, heralding the Space Age. Space exploration was no longer science fiction.

The word "orbit" acquired special significance. An orbit may be described as a path or trajectory traced by a celestial body or a satellite, under the influence of gravitational or other force to go around the earth or another body. Although natural forces basically determine the shape and size of orbits of artificial satellites, human ingenuity and imagination have described the orbits in amazingly accurate detail.

Determination of the earth's shape from the orbits of artificial satellites:

> Near-earth orbits of artificial satellites determine the shape of the earth. By 1950s it was known that the North and South Poles are nearer the earth's centre by about 21 km than is the equator. The Space Age offered a new order of accuracy and a new insight. An analysis of satellite paths around the earth in 1958 showed that the polar diameter was shorter than the equatorial diameter by 42.77 km. The flattening of the poles was confirmed but was overestimated by 170 m. In the same year, an American satellite Vanguard-1 was designed to go into a 500-km circular orbit, but it was found deviating from the predicted path.

An astronomer, John A. O'Keefe, explained the difference and said that the path was narrower on top of the earth and wider at the middle than had been believed before. It was also confirmed that the South Pole was about 40 m nearer to the centre of the earth than was the North Pole. This led scientists to say that the earth was slightly pear-shaped.

Desmond King-Hale of the Royal Aircraft Establishment (UK) refined the idea. He and his colleague, G.E. Cook, carried out a detailed study of the earth's shape, based on the paths of twenty-seven satellites. Accordingly, the South Pole was found to be 25.8 m below the level previously accepted, while at the North Pole it was located 18.9 m above the flattened sphere. In other words, the South Pole is nearer to the centre of the earth than the North Pole by 44.7 m and not 40 m. This asymmetry about the equator affects the orbits of an earth satellite by resulting in a difference of about 10 km in the orbit, which is about 200 times the discrepancy causing it. This would mean that if the orbit is measured accurately to the nearest 200 m, the shape of the earth could be known to be within about 1 m.

Equatorial Bulge

An important feature of the earth is its equatorial bulge. The radius at the equator (6,378.14 km) is longer by around 21 km than at the poles (6,356.76 km). This extra radius makes the difference for certain orbits near the earth. As a satellite goes from the southern to the northern hemisphere in west–east, its orbital plane swings westward under the influence of the gravitational pull at the equator. If the satellite goes to the northern hemisphere in east–west direction, the orbital plane swings eastward. This perturbation is known as "regression of the nodes". Nodes are the points at which a satellite crosses the equatorial plane of the earth.

Besides the shift of the nodes, the orbital plane has another apparent rotation to an observer on the earth. This is caused by the daily rotation of the earth around its axis. By the time a satellite in a low orbit completes one full orbit (say in one-and-a-half hours), the earth would have turned by 22.5° to the east. An observer will see the plane of the orbit shifted westward by the same angle.

Geodesy

Satellites have added a new dimension to geodesy, the science of surveying and mapping the earth's surface. By revealing the gravitational anomalies of the earth, they have helped scientists to outline the real shape of the earth. In addition, satellites have become accurate tools to determine location, distance, and direction which are all useful in several fields, such as navigation, geophysical prospecting, civil engineering, and intercontinental missile deployment.

Evidently, the shape of the earth is not an exact circle. Actually, it is elliptical. The shape of the earth varies with longitude as well as latitude. An analysis of orbits of a large number of satallites has revealed a new profile of the earth.

Impact of Gravity and Geoid on the Structure of the Earth

The basic reference shape of the earth's surface in terms of the gravitational equipotential, called "geoid", almost coincides with the mean sea level of oceans (ignoring the effects of tides, winds, etc.) upon which mountains and valleys are superimposed. One can find the shape of the sea-level surface, if the gravitational pull is known in detail. The extent of gravity can be deduced by the small changes in orbits that a satellite undergoes.

The geoid maps of the pear-shaped earth showed new features of the earth. NASA's Seasat, launched in 1978, gathered considerable date on the shape of the earth by monitoring the deviation in the height of the sea surface. Two Italian satellites, Lageos-1 and Lageos-2, were sent into earth orbit (1992) at an altitude of 590 km, but at different inclinations. The data, obtained by the researchers, enabled geologists to monitor the earth's crustal plate movements, measure the wobble in the earth's axis of rotation, and better understand the ocean tides.

Similarly, the Apollo-11 moon-landing mission gave new ideas on gravity measurements. A satellite, Gravity Probe-B of NASA and Stanford University, launched in April 2004 into an earth orbit of 640 km, is designed to test the extraordinary effects of gravity on space–time as predicted by Einstein.

5

The Sunsynchronous Orbit

THE Polish astronomer Nicolaus Copernicus (1473–1543) observed that the orbits of the planets were circular. He was wrong, as pointed out later by Tycho Brahe (1546–1601), who held correctly that the planets did not move in perfect circles. His famous assistant Johannes Kepler (1571–1630) was more precise. He declared that the orbits of the planets were elliptical and that as the distance of the planet from the sun increased, the planet would take longer to complete an orbit and that the planetary speed also varies with the distance from the sun.

On the other hand, a satellite's orbit around the earth is fixed along three mutually perpendicular axes: *X, Y,* and *Z*. The *X*-axis is along the line from the earth's centre towards the first point of Aries in the sky. The *Y*-axis is along the equator from the earth's centre. The *Z*-axis is along the line from the earth's centre to the North Pole. An elliptical orbit around the earth is defined by six elements and the satellite's position is accordingly monitored and corrected, wherever possible.

It is now known that a satellite speeding at 7.91 km a second will go into a circular orbit of the earth. If the satellite's speed is slightly increased, its orbit will become elliptical. In the sunsynchronous orbit remote-sensing satellites function. A sunsynchronous satellite keeps pace with the sun throughout the year. The earth's equatorial bulge facilitates the sunsynchronous orbit.

The earth's equatorial bulge pushes the orbital plane of a satellite to the east, provided the satellite passes from east to west from the earth's southern hemisphere to the northern hemisphere. The resulting eastward rotation of the plane of the satellite's orbit (precession) is equal to the apparent motion of the sun, which is about 1° eastward a day. As a result, the satellite's orbital plane always makes the same angle to the sun–earth line throughout the year, as the earth goes round the sun.

Several sunsynchronous orbits are possible. The angle of inclination of the orbital plane to the equator will vary, depending on the altitude of the satellite. For example, a satellite in a circular orbit at an altitude of 800 km should have an inclination of 98.4° if it is sunsynchronous. Satellites in orbits at about 80° inclination overfly most of the inhabited regions of the world. For example, a satellite at 893 km at an orbital inclination of 99° will make fourteen orbits a day.

The perigee and the apogee of the sunsynchronous orbit keep changing under natural disturbances and the earth's drag. If uncorrected, the perigee will vary from the nominal height, resulting in variation in the satellite's altitude over different latitudes. And it would not be possible to get the same scale of the imageries from orbit to orbit. Hence, the perigee is "frozen" by controlling its height at a specified distance by using the onboard thrusters.

In the IRS-1D mission (1997), for example, the perigee was frozen at 809 km near the North Pole. The perigee of its orbit had to be raised over a series of carefully planned manoeuvres. The satellite's initial orbit was elliptical, with its apogee at 821 km. By using onboard thrusters, the orbital velocity was increased, raising the perigee to 737 km, with the apogee remaining at 821 km.

Besides the orbital perturbation caused by the earth's oblateness, there will always be periodic disturbances caused by the residual atmospheric drag, resulting in a cumulative deformation of the shape of the orbit. The deformation can almost be nullified by selecting the initial orbit. However, air drag will reduce a satellite's orbital velocity, its altitude, and the duration of its orbit. Air drag depends on the intensity of solar activity. If it is high, the molecular density at 800 km in the atmosphere will be high. Moreover, the inclination of the orbit will also slowly decrease mainly because of the gravitational forces of the sun and the moon. Orbit corrections will be needed to compensate for the air drag as well as to counter deviations in the inclination.

The disturbances are countered by on-board attitude-and-orbit control system. Typically, a sunsynchronous satellite completes fourteen orbits a day. For circular orbits, there is a constant rotation of the normal orbit about the earth's polar axes.

6

The Geosynchronous Orbit

In 1945, the idea of a geosynchronous orbit came. An apparently stationary satellite around the earth was proposed. The noted science fiction writer, Arthur C. Clarke put forward the idea to relay radio signals from a satellite around the earth. The proposed orbit is then named after Clarke and called the Clarke Orbit.

Clarke's prophetic idea became a reality within two decades. Some British scientists tried the proposal in 1959 using the moon. Radio signals were sent to the moon, which echoed them back. After a period of 2.6 seconds, the signals were detected in a laboratory in America.

With the dawn of the Space Age, scientists tried the idea with a satellite. Communication relay equipment was placed on a satellite sent in an elliplical orbit (180–1490 km). It was launched in December 1958 and re-entered the earth in January 1959. An attempt was then made to send a metalized balloon, 30 m across, in circular orbit of the earth and it sent back a signal from space. It was called Echo and when launched in 1960, it performed well. It reflected radio waves as it circled the earth every 100 minutes. However, because of its low orbit, it passed over the ground stations rapidly. As it did not amplify the signals, it was called a passive satellite and it needed powerful ground transmitters to catch its weak signals.

The next step was to develop an active satellite, which amplified the signals sent from the earth. In 1962, a satellite called Telstar transmitted television signals from one continent to another. The transmission between America and France lasted 22 minutes before the satellite went out of sight. As the satellite passed over quickly, it was realized that such satellites should be made to orbit at a greater height so that their speed could be decreased. In devising Echo and Telstar, John Pierce of the Bell Laboratory (USA) played a key role.

In February 1963, Syncom-I, the first satellite in synchronous orbit, was launched, but it lost contact with the earth following a malfunction. Syncom-II followed on 26 July and became successful. It was the first geosynchronous satellite. The plan was outlined by Harold A. Rosen of Hughes Aircraft Company, USA. The satellite performed well. In 1964, Syncom-III over the pacific relayed the Tokyo Olympics to America. Soon, international and domestic agencies realized the great potential in the use of geosynchronous satellites.

Characteristics of a Geosynchronous Orbit

If a satellite is orbited at 11,068.56 km/hr (or 3.0746 km/sec) at about 36,000 km (typically 35,786 km to be exact) above the equator, the satellite will synchronize with the speed at which the earth rotates on its axis at the equator (1,620 km/hr or 464 m/sec). Further, the satellite should orbit from west to east, as does the earth. As the satellite's speed is synchronous with that of the earth, it is called a "geosynchronous satellite".

The orbit is called "geostationary", if in addition to it being geosynchronous, the plane of the satellite's orbit coincides with the earth's equatorial plane (at nearly 0° inclination), without any change in its shape. The satellite would then appear as a fixed star to an observer on the earth.

The satellite's orbital period is one sidereal day, which is the time taken by the earth to rotate once relative to the stars. It is a bit shorter (23 hours, 56 minutes, and 04.09053 seconds) than a solar day of 24 hours, which is the duration of the earth's rotation relative to the sun. This is because by the time the earth has rotated once, it has also moved a little in its orbit around the sun and so the earth has to rotate for about another 4 minutes before the sun appears in the same place in the sky as it did a day before. In each orbit, the satellite covers a circumference of 265,000 km.

A geosynchronous orbit will be a circle at a uniform distance of 42,164.2 km from the centre of the earth, with the earth's equatorial radius of 6,378 km and the distance of the satellite from the edge of the earth, viz. 35,786 km.

7

India's Missile Mission

THE foundation of India's missile mission was laid by Dr A.P.J. Abdul Kalam, the architect of the missile programme. All the strategic and tactical missiles are made in the country.

In 1983, the Integrated Guided Missile Development Programme was taken up by the Defence Research and Development Organisation (DRDO) of the Ministry of Defence. The objective was to design, develop, and lead the production of four missile systems, viz. Prithvi, Trishul, Akash and Agni. A recent addition is BrahMos (derived from Brahmaputra and Moscow), being made with Russian collaboration. The armed forces will be provided the missiles, which will have the state-of-the-art weapon capability.

India has developed missiles like Prithvi, Agni, Akash, Trishul, Nag and BrahMos. Let us discuss about the preparation and performance of these missiles.

Prithvi

India's Prithvi missile is a short-range surface-to-surface battlefield missile. It has a single stage liquid propulsion system. The Army version of the missile has a range of 150 km and it can take a 1-tonne playload. It has inertial guidance and onboard computer to ensure the desired accuracy on impact. Warheads can be interchanged on the field and the missile's trajetory is manoeuvrable. The Air

Force version with 250-km range has undergone flight tests. A Navy version, called Dhanush, which has a range of 350 km, had been test-fired. Another version, Prithvi-III, can be fired from the ground vehicle or ship.

Agni

It is an intermediate-range ballistic missile. Agni-I has a solid propellant booster and a liquid propulsion upper stage. It has a short range of 800–900 km. Agni-II is a two-stage, all-solid propellant system. It has a range of over 2,500 km, which can be reached in about 11 minutes. It is 20 m long and weighs 16 tonnes and has mobile launch capability. It has a carbon composite structure for protecting the payload during the re-entry into the atmosphere. Its re-entry system can withstand a temperature of 3000 °C at the hypersonic speed of 12-14 Mach number. It needs a super computer for flow-field analysis and advanced radar for multi-target tracking. The missile can carry "special weapons" payload of over 1,000 kg. Due to improvements of its components, it can attain specific targets precisely. Agni-III, a surface-to-surface missile, has been developed with a range of 3,500 km.

Akash

Akash is a medium-range surface-to-air missile with multi-target engagement capability. It can track sixty-four targets simultaneously. It uses high-energy solid propellant for the booster and ram-rocket propulsion for the sustainer phase. Initially commanded, it operates later under radar homing guidance with high accuracy. It can deploy missiles to attack four targets at the same time within a range of 25 km. The system is highly mobile. It can down high-flying airplanes.

Trishul

It is a quick reaction surface-to-air missile, which can be fired from a moving vehicle. It can also be used as an anti-sea skimmer from a ship against low-flying attacking missiles. It can counter all known aircraft jammers. It uses high energy solid propellants and is commanded initially from the ground and later guided by a tracking radar. It is a boost to mobile armoured columns. Its high manoeuvrability and high lethal capability would be useful for the armed services.

Nag

Nag is a third generation anti-tank missile system, which provides "fire and forget" and "top attack" capability. It can destroy battle-tanks. It has advanced imaging infrared guidance system and millimetric wave radar, eliminating the need for human direction. Nag can be fired from a tracked vehicle or a helicopter. It is ahead of contemporary systems under development in the West. Several flight tests have been carried out.

BrahMos

It is a supersonic missile. Only Russia and India have such missiles; not even the US has it. This missile can attain 2.8-3 Mach speed at a height of 15 km. It can be launched from anywhere — silos, mobile platforms, ships, submarines, and aircrafts with no design changes. The first flight was conducted in 2001. In subsequent tests, the missile has been launched from various platforms, except from submarines and aircraft. It can deliver a warhead of 200–300 kg within a range of about 300 km. Though the missile is essentially for anti-ship operations, BrahMos-II has been designed for land-to-land deployment. It is a two-stage missile, with a solid propellant booster and a liquid propellant ram jet system, with anti-jamming devices. An Army version can be fired from a mobile launching complex from the ground. New versions for the Army are being designed.

An air-launched version of BrahMos has been planned to be developed by India and Russia. BrahMos is the most competitive missile system in the world. It can be flown both horizontally and vertically. An Indian–Russian joint venture, BrahMos Aerospace, has been formed.

8

Indian Rockets

Prelude

India became free from British rule in 1947 and got Independence. Then Indian scientists tried to uplift the country. A young Indian scientist, Vikram Sarabhai (1919-71), founded a research laboratory for the study of cosmic rays. He became interested in this field of research under the influence of Nobel laureate, C.V. Raman (1888–1970). Another famous scientist, Homi Bhabha (1909-66), also studied cosmic rays in the 1940s. Sarabhai realized the potential of satellites for developing countries and worked to set up the basic institutions necessary for undertaking a space programme. The Indian National Committee for Space Research was formed in 1962 to identify the country's goals and priorities in the area.

Sounding Rockets

Launching of sounding objects was started from Thumba near Thiruvananthapuram, in view of its location near the magnetic equator. In 1963, Thumba became operational with the launch of a foreign rocket to study the atmosphere near the magnetic equator. Several nations launched sounding rockets from Thumba.

The Indian Space Research Organisation (ISRO) implemented a programme of making sounding rockets and propellants for them.

Necessary developments were made at Thumba for launching the rockets. A Space Science Technology Centre was also set up at Thumba in order to carry out related investigation programmes. With a view to taking advantage of a launch centre on the east coast, the island of Sriharikota was chosen and developed as a launch centre.

Then, a space centre was established in Ahmedabad to provide the necessary thrust for application of space technology in various fields of development. In 1972, a satellite project was estabished in Bangalore, which was subsequently developed as the ISRO Satellite Centre responsible for the design, development, and launching of all spacecrafts. Beside the lauching centres at Sriharikota and Thumba, Candipur in Balasore, Odisha was developed as a launching centre, with the main control centre in Bangalore. The tracking and command network includes, besides the launch ranges, centres in Lucknow, Car Nicobar islands, and Mauritius. There is also an optical satellite tracking and ranging station at Kavalur, Tamil Nadu.

Rohini-75

The first Indian rocket, called Rohini-75 (the number indicates its diameter in mm), was launched from Thumba in 1967. This success led to make a two-stage rocket called Centaur and the engineers started regular production of sounding rockets in 1971. The Rohini series expanded. The diameter of the rockets steadily increased along with their improvements.

Rohini-100, RH-125, RH-300 and RH-560 were soon launched, for carrying payloads to study the atmosphere and for testing new subsystems for bigger launch vehicles. The new progress made in sounding rockets led to the design and development of a Satellite Launch Vehicle (SLV).

Over 1,000 sounding rockets including those made in the country have been launched. Starting from a single-stage rocket weighing 100 kg, four-stage rockets weighing 17,000 kg each have been made and launched. From unguided to guided rockets, considerable progress was made in control and guidance as well as telemetry and telecommand based on computers. India became the sixth nation in the world to have achieved satellite-launching capacity.

SLV and ASLV Programmes

Failure is the pillar of success. The first SLV-3 rocket launched from Sriharikota Range (SHAR), now Satish Dhawan Space Centre ended in a failure in August 1979. However, the second attempt was successful. In July 1980, a Rohini satellite (35 kg) was orbited by SLV-3. The second development flight (SLV-3, D2) became success. It put a 41.5 kg satellite into a near-earth orbit of about 400 km × 850 km. Two more launches of SLV-3 were conducted in May 1981 and April 1983.

ISRO subsequently started work on an Augmented Satellite Launch Vehicle (ASLV), designed to put a 150 kg satellite into near-earth orbit. Two strapon boosters were added to the first stage to provide extra thrust. However, the first development flight in 1987 and the second in 1988 did not realize the mission. The first launch failed to ignite in 1987 though the computer had given the command.

ASLV-D2 was launched in July 1988 after incorporating several improvements suggested in the wake of the failure of the first ASLV. The second attempt also failed. With various technical modifications ASLV-D3 was made. Hundreds of test runs were conducted on a computer and simulations were carried out. For the first time, the indigenously developed fuel, hydroxyl-terminated polybutadiene (HTPB), was used for the core first stage and strapons in place of Polybutadiene Acrylic Nitrile (PBAN).

In May 1992, ASLV-D3 was launched and it became successful. The objective was to operationalize the ASLV's capability of placing a 150 kg satellite — as part of ISRO Stretched Rohini Satellite Series (SROSS) — into a 400 km near circular orbit and to validate new technologies for future launch vehicles and satellite programmes.

The launcher could not, however, place the satellite in the desired orbit. Its fourth stage did not generate the required spin to the satellite. Hence, the orbit achieved by the satellite measured only 433 × 267 km.

The problem was solved in the next mission, ASLV–D4, in May 1994. The spin-up system of the combined stage and the satellite imparted the required spin rate (140 rpm). After the injection of the satellite into orbit and the separation from the fourth-stage, the spin rate was reduced by the satellite system. And the power supply to the pyrotechnic ignition was increased. The payload weight was reduced from 150 kg to 106 kg. Additional fins were inserted to increase the stability of the first stage. Closed loop guidance was introduced from the ignition of the second stage until the separation of the third stage.

The closed loop guidance accurately guided the vehicle by providing real-time dicisions on-stage ingnition and separations. The events were not pre-fixed as was done in SLV-3 and the first two flights of ASLV, but were set by the on board real time decision system, based on a comparison of the actual motor pressures and acceleration with pre-set values.

The modifications became successful. The longevity of the satellite in the desired orbit increased. The satellite (SROSS-C_2) was successfully manoeuvred into the final intended orbit of 429 × 628 km for performing scientific experiments.

Further improvements to enhance the reliability of solid boosters in India became possible. India has grown the solid rocket technology. Indian solid rockets can compete in the international market for launching small satellites in low-earth orbit.

9

GSLV: A World-Class Rocket

THE Geosynchronous Satellite Launch Vehicle (GSLV) developed by ISRO is a three-stage vehicle. The first stage has a solid propellant (129 tonnes) core motor, with four strapons, each with 40 tonnes of storable liquid propellants (instead of six solid propellants strapons of a PSLV). The second stage has storable liquid propellant of 17.5 tonnes, derived from the second stage of a Polar Satellite Launch Vehicle (PSLV). A single cryogenic third stage replaces the third and fourth stages of a PSLV and has 12.5 tonnes of propellant (liquid oxygen and liquid hydrogen), with 7.5 tonnes of thrust in vacuum.

Launching of GSLV

On 28 March 2001, the first developmental flight of GSLV was attempted. Due to some technical error, the launch could not be successful initially. It was soon detected that defective plumbing in the oxidizer flow of the engine was the cause of the flow thrust. The technical snag was corrected within a short period. The defective strapon engine was replaced. The launch took place on 19 April 2001. It became successful. However, the satellite's orbit was less than the desired value. The height of the highest point of the satellite it carried (GSAT-1) was less by 3,800 km. It was later traced to under-performance of the third stage. Further developments were done.

GSLV's experimental satellite, GSAT-1 weighing 1,540 kg, was designed to prove new elements such as fast recovery star sensors and 10 Newton reaction control thrusters. It had two C-band transponders for demonstrating digital audio broadcast, Internet services, and compressed digital TV experiment.

The second development flight of GSLV, with GSAT-2 onboard weighing 1,825 kg, was launched successfully in 2003. The additional weight was possible because of the enhanced propellant loading in the core of the first stage and use of a high-power engine in liquid propellant strapons as well as in the second stage, besides some structural improvements.

First Operational Flight

The rocket placed the geosynchronous satellite, EDUSAT, in the correct transfer orbit. Due to the prediction of rain in Chennai, ISRO did not want to take chances. It sent up a balloon to verify the winds in the upper atmosphere at about 20 km. A radar developed by ISRO was also used. The trajectory was fine-tuned in accordance with the wind data. The strategy was called zero angle of attack, which simply means avoiding undue resistance to wind forces.

After all obervations regarding the weather conditions, the rocket was cleared for launch at half past nine in the morning. The countdown did not have a single hold. Still, as the countdown was nearing its end, tension mounted. But, every performance went good. Finally, 138-tonne first stage was ignited and it rose majestically above the ground.

As a rocket's thrust at ground level should be more than the total mass of the rocket including the payload, the first stage and its four strapons together produced 7,952 kiloNewton thrust to lift a weight of 414 tonnes, resulting in a high thrust-to-weight ratio.

After the first, second, and third stage of the process of firing, the rocket acquired the required velocity. At 1,014 seconds from the start, EDUSAT was put into orbit. It was then 5,000 km away from Sriharikota, over Indonesia. The separated cryogenic stage was reoriented so as to avoid any collision with the satellite.

The velocity of the satellite at injection into the orbit was 10.2 km a second as planned. If it were 11 km a second, the orbit would have been an ellipse and at 7.9 km the orbit would have become circular. The satellite's solar panels were deployed shortly after it was in orbit. All the sequences went off well as planned. The satellite was monitored by the ground stations at Biak in Indonesia (set up by ISRO), Canada, Italy, and Beijing, besides the Master Control Facility (MCF) at Hassan (Karnataka) and the ISRO tracking network in Bangalore.

The launch was a grand success. There are some merits of GSLV. Not all tests can be made on the ground. Still as many as 3,000 simulations are done for a GSLV. Also, three-axis sensors ensure accuracy of the rocket's path along the pitch, roll-and-yaw directions within a fraction of a second. There is also what is known as the "event management" computer system onboard for correcting deviations.

The egg-shaped transfer orbit which EDUSAT was injected had a perigee of 180 km and an apogee of 35,985 km. Subsequently, the MCF at Hassan raised the perigee in stages by firing the onboard apogee boost motor on three days for a total of 94 minutes after precise calculations, to add an additional velocity of 1.66 km a second and make the orbit circular and near-geosynchronous with almost zero inclination to the equator. The satellite drifted towards its allotted position at 74° East longitude to be co-located on 2 October 2004, with Kalpana and INSAT-3C. At the height of about 36,000 km above the equator, the satellite appears as a fixed star, as its speed synchronizes with that of the earth. The satellite's three-axis stability system was switched on, with momentum wheels running at 4,500 revolutions a minute. The satellite is likely to function throughout its designed life of seven years, as it is left with 300 kg of fuel (out of 1,128 kg, it had at the beginning of the orbit), adequate for corrections during the period.

In terms of thrust, the first stage develops 4,700 kiloNewton for 100 seconds. Each of the four strapons as well as the second stage has 680 to 700 kiloNewton thrust. The strapon burns for about 160 seconds while the second stage burns for 150 seconds. The third stage produces thrust of 75 kiloNewton for about 750 seconds. Even as ISRO uses the Russian engine and the stage, its indigenous versions are getting ready.

New features of GSLV include an improved inter-stage between the first and the second stages. It will facilitate the firing of the second stage even as the first stage completes its thrusting action.

The control and guidance, telemetry and telecommand are placed above the equipment bay. A heat shield (larger than in PSLV) protects the vehicle's electronics and the spacecraft during the ascent through the atmosphere.

The rocket has a three-axis attitude control stabilization. It is achieved by autonomous control systems in each stage. The inertial navigation and guidance system in the equipment bay guides the vehicle from the lift-off to injection into orbit. The digital autopilot and closed loop guidance system on board ensure the required altitude change and guide the injection of the satellite in the specified orbit.

GSLV has hundreds of subsystems, which are designed, manufactured, tested and qualified before integration for launch. About 150 industries, both public and private, are involved in the supply of materials. However, not all systems can be tested on the ground. Quite a few can be tested and evaluated only in space.

The Failure of GSLV-F02

The launching of GSLV-F02 in July 2006 became a failure. It crashed into the Bay of Bengal about 60 seconds after its take-off. The

Failure Analysis Committee searched for the reasons and arrived at a conclusion that the basic design of the GSLV is robust and suggested that the automatic launch sequence be strengthened.

In both GSLV and Falcon mission, the damage was caused by relatively minor components, though the systems are perfect. It only shows that nothing is too small in space affair.

10

Prospects of INSAT System

INSAT-1

A THREE-IN-ONE satellite is called INSAT, which combines the playloads for telecommunications, television, and meteorology instead of having one satellite for each of these services. The Government of India decided to buy the first operational satellite from abroad. An American company, Ford Aerospace and Communications Corporation, won the contract for providing the satellite.

The satellite had a box-type structure with a deployed solar sail, a solar array, and different antennas for telecommunication and TV transmission. The satellite had one big and twelve small thrusters to be used when needed for keeping it steady.

Failure of INSAT-1A

INSAT-1A was launched on 10 April 1982 from Cape Canaveral by a Delta-3910 rocket. The first major problem occurred early in the mission. The communications antenna did not open initially. The next major problem was with the solar sail, which did not open. After corrections of these defects, the operational services were started. However, on 4 September 1983, as the altitude control system was being operated on a routine mode, the sun and the moon intruded on its sensors, resulting in the loss of contact with the earth. Although the Master Control Facility (MCF) immediately started emergency procedures, critical commands did not reach the satellite and the status of the spacecraft's systems including the fuel level did not reach MCF. As a result, the fuel got depleted even before anyone could know about it and the spacecraft was lost.

INSAT-1B

In September 1983, INSAT-1B was launched from a US space shuttle. When it was put in the transfar orbit, the apogee motor was fired on command from the ground. The firing was 9 minutes shorter than planned. The motor was again fired, but this time it was a bit excessive. The result was later corrected. The first TV picture was received on 15 September but it was only a month later (15 October 1983) that the satellite was brought to its slot.

INSAT-1B functioned well for over seven years, longer than its design life. In order to augment the INSAT capacity and provide back-up services, transponders were taken on lease from INTELSAT, ARABSAT, and INTERSPUTNIK.

INSAT-1C

It was launched in July 1988 by an Ariane rocket. After the satellite was placed in the orbit, it experienced a short circuit in the power system, and only half the total power of the satellite was made available. Hence, half the communication and television channels were operated. Only the meteorological package worked fully. On 22 November 1989, the spacecraft lost contact with the earth.

INSAT-1D

INSAT-1D was scheduled to be launched by a shuttle in mid-1987. A shuttle could not be made available. It was finally launched by a Delta-4925 in June 1990. All services of INSAT-1B were transferred to INSAT-1D.

Second Generation of INSATs

The second generation of INSATs was indigenously developed and fabricated. INSAT-2A was launched on 9 July 1992, followed by INSAT-2B on July 1993. Several new features were introduced. Latest designs were used for such systems. Basically, the satellites ensured continuity of services.

In order to cover remote areas, there was a large increase in the number of telecommunication earth stations. A large number of Very Small Aperture Terminals (VSATs) have emerged to form data networks. Regional TV services have increased.

Each of the INSAT-2 satellites had eighteen transponders in the C-band, including six extended C-band (4,200-4,500 MHz and 6,425-7,025 MHz). The latter has several advantages, such as interference-free communication and suitability to Indian conditions. In addition, each INSAT-2 had a separate transponder for weather (Very High Resolution Radiometer (VHRR) — one for data relay and two for television and other services.

INSAT-2C

Two more satellites in the INSAT-2 series, INSAT-2C, and INSAT-2D were designed without the weather monitoring payload, since there was ample provision for it in the satellites working in 1995. As a result, the shape of the latter two satellites was different, with polar panels on either side.

The payload provided for fixed satellite services, including for the first time three Ku-band (11-14 GHz) transponders; broadcast satellite services and mobile satellite services. Higher frequencies in the Ku-band were designed for smaller receiving dishes and for serving a large number of VSATs. The satellite covered a wider region from West Asia to Southeast Asia. Mobile communication from the geostationary orbit was attempted. The power available on board increased to 1600 W (as against 1000 W in INSAT-2A and INSAT-2B) with the introduction of two-sided solar arrays.

INSAT-2D and INSAT-2E

In 1997, INSAT-2D was launched in June. It worked only for about four months. After improvement, the satellite INSAT-2E was launched in April 1999 from Kourou, French Guiana, by an Ariane rocket. INTELSAT leased part of the capacity of INSAT-2E. The indigenous INSAT-2E has seventeen transponders. This satellite weighs 2.6 tonnes. It has better thermal control components, improved solar cells and high-capacity batteries. The satellite has a mission life of twelve years.

Each INSAT-2E transponder can accommodate two TV channels, instead of one as in the earlier satellites. Several Indian broadcasters (e.g. Enadu TV, Sun TV and Asianet) have switched over to INSAT-2E.

The commissioning of INTELSAT-2E has strengthened the telecom services and TV broadcasting. The satellite also carries an advanced meteorological payload.

The third generation INSAT series began with the launch of INSAT in March 2000 by Ariane-5 from Kourou. It is a communication satellite launched ahead of INSAT-3A and has twelve extended C-band transponders needed to meet the demands, following the failure of INSAT-2D. This was followed by INSAT-3C in 2002, and INSAT-3A and INSAT-3E in 2003. INSAT-3D is a weather satellite, launched later.

The INSAT-4 Series

INSAT-4 series began in December 2005. An Ariane rocket launched INSAT-4A from Kourou. It is ISRO's most sophisticated satellite with twelve Ku-band transponders for providing direct-to-home television in the country and another twelve transponders in the C-band for communications within the country and beyond in the south-east and north-west regions.

INSAT-4A had a lift-off mass of 3,080 kg with a dry mass of 1,987 kg. Its solar arrays can provide 5,500 watts of power. This satellite has a design life of twelve years. India plans to launch ten-twelve communication satellites by 2010. Six more satellites in the series are underway.

11

Ariane and Other Rockets

BESIDES USA and Russia, China, Japan and the European Space Agency (ESA) have developed their own rockets to launch satellites. Other countries can pay and use their rockets (expect those of Japan) for launching their satellites.

Ariane is the European rocket. It is named after the Cretan Princess, who helped the Greek hero, Theseus, escape from a labyrinth. It caters to half the world market in launch services for civilian space flights. Ariane rockets have launched many INSATs. In its early days of operation, Ariane carried India's APPLE (1981).

The first Ariane rocket was launched on 24 December 1979. It is now an established commercial launcher, capturing over 50 per cent of the worldwide launcher market. Ariane had placed 160 satellites in 121 launches until October 1999.

Increasing the mass of the satellites, Ariane-2 and Ariane-3 were flown in 1984. These were succeeded by Ariane-4 in 1988, which was available in six versions with two or four solid or liquid strapon boosters attached to two liquid propellant and one cryogenic stages.

Ariane-1 had a Geostationary Transfer Orbit (GTO) capability of 1,835 kg, Ariane-2 and Ariane-3 increased it to 2,275 kg and 2,650 kg, respectively, Ariane-4 with four liquid boosters allowed dual launches of 1,140 kg or 2,500 kg. It could also place satellites weighing 4,000 kg (e.g. INTELSAT-VI and INTELSAT-VII). The standard deviation in GTO was only 0.5 km for the perigee and 30 km for the apogee.

Ariane-5 is a more powerful and more reliable launcher. This rocket has a capacity to place a satellite weighing 5,970–7,400 kg in GTO or 23 tonnes in low-earth orbit. The cryogenic motor is ignited and controlled before take-off.

The first two flights of Ariane-5 in 1996 and 1997 failed. However, the third qualification flight in October 1998 was a success. The launch of Ariane-5CA, from Kourou on 12 February 2005 was a complete success.

The 780-tonne Ariane-5CA placed two satellites correctly in the orbit in about 30 minutes after the lift-off. It has proved its capacity to put 10 tonnes into space. Specifically, it can place a 9.6-tonne satellite into GTO.

American Rockets

The United States has 413 known satellites launched by a variety of rockets, such as Scout, Thor-Agena, Atlas-Agena, Atlas-Centaur, Gem-Titan, Saturn I, Uprated Saturn I and Saturn V. Besides, satellites like INTELSAT-V and Mariner Mars Orbiters are in the line.

The Saturn V, America's most powerful staged rocket, was used to send astronauts to the moon. Its first manned flight, Apollo-8, was also the first manned flight to the moon, though it did not land on it. Apollo-11, launched by Saturn V, in 1969, achieved the first lunar landing. The rocket also lifted the unmanned Skylab into the earth orbit. Saturn-1B launched three manned missions to the Skylab and later to the Apollo-Soyuz docking mission.

Delta was derived from Thor intermediate range ballistic missile in the mid-1950s. In 1960 NASA successfully launched a satellite Echo with a Delta. The Delta rocket has IIIrd and IVth versions. A new set of Delta-IV and Atlas-V (with Russian engine) launchers mission has been developed.

Russian Launchers

Russia has launched the world's largest number of rockets. Since the early 1960s, it has made the Proton rocket (in three-or-four-stage versions) without essential changes in the design, but continued it as one of the most reliable launchers in the world. It has launched a variety of satellites in the earth orbit, as well as in interplanetary missions. Proton can deliver a payload directly to a specified point of the Geostationary Orbit (GEO), without the need for an artificial apogee motor on artificial satellites. In a new version (Proton-M), the control systems would be upgraded.

Russia developed an oxygen-hydrogen booster, called "Breeze-M", which can be used for payload delivery to near-earth orbits and middle altitude orbits. It can also be combined with many other launchers such as Zenit and Ariane-5. Another type of rocket, called "Angara" has been designed, which is capable of launching up to 30 tonnes close to the earth. A Proton-LV "Zarya" was launched in 1998 which was the first element of the International Space Station.

Russia launched Soyuz rockets. The Soyuz rocket has launched 1,696 manned and unmanned spacecraft since 1963 as of May 2005, accounting for about half of all Russian satellite missions.

Chinese Rockets

In 1970, China launched its first rocket, called Chang Zheng, meaning long march. Then, the country has developed a series of long march launch vehicles to place a variety of satellites in different kinds of orbits, viz. LEO, GTO, and SSO. China can rival the powerful Western boosters, with a GTO capacity of 5,000 kg. It has developed a series of launch vehicles for low-earth orbit missions with a capacity of 300-9,200 kg.

After the rocket, China launched Long March-2 (LM-2C) and then LM-2D. The next version, LM-2E was also launched. In 1990, it launched a Low Earth Orbit (LEO) satellite for Pakistan. LM-2E launcher was a failure in 1995 due to the explosion of a US-built APSTAR-2 satellite.

The LM-3 series was for Geosynchronous Transfer Orbit (GTO). The three-stage LM-3 was an all-liquid rocket with a cryogenic third stage, capable of placing 1,500 kg in GTO. LM-3A could place 2,700 kg, while LM-3B with four boosters was the most powerful in the series, with a capacity of 5,000 kg in GTO. The next version LM-4, also a three-stage vehicle, could be used for polar and sunsynchronous missions. LM-4A has an improved third stage.

In 1999, China inaugurated LM-4B, an all-liquid three-stage vehicle with a payload capacity of 2,800 kg for a 900 km sunsynchronous the orbit. It launched a weather satellite and a scientific satellite into the polar orbit. An LM-4B launched a joint Chinese–Brazilian Earth Resources Satellite (1,450 kg) in the sunsynchronous orbit.

China began a manned space programme in 2002. An experimental spacecraft, Shenzhou-1 was launched. After four such missions, the country launched Shenzhou-5, which took an astronaut. Yang Liwei completed fourteen orbits around the earth at about 343 km and descended in good shape.

In the Shenzhou-6 mission in 2005, Fei Junlong and Nie Haisheng were in the orbit for five days. They parachuted to safety and landed in inner Mongolia. Shenzhou was bigger and had more power than Soyuz.

Japanese Rockets

Japan sent its first rocket into space in 1970. Its four-stage all-solid propellant rocket, Lambda-45, was launched by the Institute of Space and Astronautical Science. The National Space Development Agency (NASDA) then took over with a sustained programme. NASDA launched a three-stage N-I, incorporating US technology. In 1972, N-I, with a liquid second stage, placed a 130 kg satellite in GEO.

N-I boosted seven satellites into the orbit by 1982. The next vehicle, N-II with three stages and strapon motor had thrice the thrust of N-I and placed eight satellites into GEO between 1981 and

1987. A cryogenic stage was introduced in the next generation of rockets, called H-I. It could place 550 kg satellites in GEO. Eight satellites were launched without a failure in 1986. A fully indigenous rocket H-II was launched in 1994. The second H-II launched in 1994 encountered problems.

Due to several launch failures of H-II, Japan abandoned the project. In February 2005, Japan launched an H2A rocket, which can be configured with a mix of solid or liquid boosters to take up 7.5 tonnes of payload to GTO near the earth.

Launching a Rocket From a Ship on Sea

With the advantages of technology, sea launch came to a reality. A ship with a platform is used to launch satellites from the equator (0° N latitude and 154° W longitude), taking the help of the earth's eastward rotational force at the site. A rocket can be launched to all inclinations from a single launchpad on board the ship. The ship platform, called Odyssey (200 m × 32 m) is a joint venture of companies in the US, Norway, Russia and Ukraine. It has a vehicle assembly area below its decks and can accommodate 240-launch crew.

In May 1999, a three-stage Zenit-3SL rocket lifted off from the platform. The rocket placed a direct broadcast satellite for service to all the fifty US states into a precisely planned GTO with an orbital inclination of just 0.53°.

In April 2005, a Zenit-3SL placed a broadcast satellite, Spaceway-F1 for direct TV into a correct orbit. Weighing over a 6,000 kg, it was the heaviest commercial satellite ever launched. The US has also made a new hydrocarbon booster engine, the first to be flown in forty years. Powered by liquid oxygen and kerosene, it is designed to accurate to twenty-five times the speed of sound in 10 minutes.

12

Ariane Passenger Payload Experiment (APPLE)

Even before the INSAT was built abroad, Indian engineers acquired first-hand experience in designing, building, launching and operating an advance geosynchronous satellite under the APPLE. It was a three-axis stabilized, experimental communications satellite launched on the third developmental flight of Ariane of ESA. The facility was offered free of charge by European Space Agency (ESA). The design and development of the spacecraft was completed in thirty-six months — a record by international standards. The 673-kg APPLE satellite called for the designing and building of an apogee motor, orbit control system and payloads for telecommunication and television experiments.

APPLE was launched from Kourou in French Guiana on 19 June 1981. The spacecraft took a position above 102° E longitude, over the equator on 16 July 1981. All mission objectives were fulfilled. It ended its effective life on 19 September 1983, a day after INSAT-1B was placed in its space home. APPLE functioned for twenty-seven months after its launching, which was well beyond its nominal design life of two years. It worked till the last drop of its onboard fuel.

A wide range of experiments was conducted, involving advanced telecommunication techniques and use of small earth stations. Television links between different cities and regions became possible through the satellite. For the first time, national events could be telecast simultaneously to various parts of the country. The satellite link was found useful in expediting cyclone relief. The space link was tried to make interbank transactions between Mumbai and Ahmedabad. A newspaper printed its remote edition through the satellite. Computer interlinks and data transmission among Ahmedabad, Mumbai, and Sriharikota were also successfully accomplished. A video course on satellite communication to postgraduate students at several centres was beamed in a teleconference mode. A video course in robotics (sophisticated technology relating to robotics) was relayed to about 500 scientists and engineers in Ahmedabad, Bengaluru, Chennai, Hyderabad and New Delhi. The pre-recorded course was sponsored by the Institute of Electrical and Electronics Engineering (USA). Some of the results obtained in an earlier project (the Satellite Instructional Television Experiment (SITE) for direct telecast to villages by an American spacecraft were checked again.

13

EDUSAT and NANOSAT

SATELLITE links are needed for imparting distance education in our country. The Indira Gandhi National Open University (IGNOU), for example, offers a number of courses under its distance education programme through satellite links. Such EDUSAT would go a long way to design and fabricate a NANOSAT, which has components on the nanoscale (one billionth of a metre).

EDUSAT

In order to support and improve the quality of education, ISRO has emphasized the use of satellites for education. The shortage of good teachers in almost all fields is fulfilled by such an arrangement.

In 2004, ISRO launched EDUSAT, a satellite exclusively dedicated for education. It would provide multimedia, interactive lessons, and will augment distance education. The satellite is specially designed for providing five spot beams in the Ku-band (that gives strong signals to small antennas on the ground) to as many specified regions, viz. east, west, south, north, and north-east, besides one Ku-band national beam for the Indian mainland. The satellite has a multiple spot beam antenna with a 1.2 m reflector to precisely direct the spot beams to the intended regions. About 100–200 classrooms can be connected in each beam.

Besides supporting formal education, the satellite will be used for providing knowledge on health, hygiene, etc. among the people in the rural and remote areas. The satellite will enable educational planners to overcome the shortage of good teachers at all levels by making their services available to a wider student population.

The satellite had a lift-off mass of 1,950 kg and a dry mass of 820 kg. Its mission life was estimated to be a minimum of seven years. It is a three-axis body stabilized in orbit using earth sensors, momentum/reaction wheels, sixteen thrusters and magnetic torques.

Visvesvaraya Technological University (VTU) has two satellite channels — the first in the country. The success of the project would depend not only on the effective functioning of the satellite, but also on the active participation of a number of agencies, for the ground realities in the education sector pose quite a challenge.

Telemedicine through INSAT

INSAT plays an important role in the medical sector. Telemedicine has now become possible. A pilot project in Karnataka began operation in 2002. Narayana Hrudalaya, a super-speciality hospital in Bengaluru and Vivekananda Memorial Hospital at Saragur in Karnataka, are linked to an INSAT satellite.

More remote hospitals and speciality centres in the country have been brought into the network. ISRO has, on its part, perfected the digital connectivity, which converts the medical images of the patients into a digital signal to be beamed to an INSAT for relay to the super-speciality hospitals. The video link is stable and clear, and the interactive mode is effective.

Elsewhere, ISRO has brought telemedicine to soldiers in Kargil, pilgrims to Sabarimala as well as the people in Andaman and Nicobar, who got the benefit the very next day after the tsunami. Telemedicine links are working in Odisha, Assam, Tripura, Ladakh, Lakshadweep, Andaman and Nicobar islands, besides Karnataka.

Internet, TV and Radio Links

Satellites are useful for Internet, TV and radio communications. Satellite television has become popular. In urban areas, cable TV is provided by private operators. The INSAT system is offered to non-governmental users on commercial terms. The INSAT system can serve over 85 per cent of the country's population. Doordarshan offers separate channels for all major languages as well as an international channel, primarily for Aisa. In addition, Direct-to-Home (DTH) service is now provided, beaming channels of Doordarshan and numerous private TV companies. The format and frequency of TV news from satellites have undergone a drastic change.

INSATs are deployed for radio networking. Over 200 radio stations receive satellite signals uplinked from select stations. The National Informatic Centre (NIC) operates a satellite-based computer communication network for dissemination of news and information to people on a variety of subjects. A standard time and frequency signal dissemination service from satellite is operated through the radio network.

ANUSAT and Micro Satellites

Micro satellites are also working. Anna University's ANUSAT, a 50-kg satellite is the forerunner. It was made by the students and teachers of the university under the guidance of ISRO.

A PSLV will launch it as a piggyback payload. Weighing around 60 kg, ANUSAT has body-mounted solar panels (for producing 40 W) and is spin-stabilized. It has a data store-and-forward payload for conducting experiments on message transfer across the country.

A PSLV can carry two micro satellites with a mass of 120 kg as piggyback payload in one launch. Four mini and six micro satellites can be launched, if a suitable platform is developed for insertion into a PSLV. Already, a PSLV has launched foreign satellites KITSAT-3 (110 kg), DLR-TUBSAT (45 kg), BIRD (92 kg),

PROBA (110 kg), satellites of Singapore University, and Italy will also be launched.

A popular micro satellite is HAMSAT (40 kg), launched with Cartosat-1 in 2005. It is the smallest satellite built by ISRO in recent times. A ham enthusiast (amateur radio operator) in the Netherlands has supplied one of the two transponders. Indian hams have also been involved in testing it. This has become the most popular amateur radio satellite at home and abroad for the relay of voice signals for hams.

Hamsat spins at about four revolutions a minute and maintains stability. As a reliable relay station in the space, it enables hams to communicate across vast distances. Some of the areas of application of micro satellites are communication from low-earth orbit, remote sensing and science experiments. In USA and Europe, several universities are engaged in developing micro satellites. The University of Surrey in the UK has made several micro satellites for disaster monitoring in several countries. The German Aerospace Reserch Centre (DLR) developed a micro satellite for fire direction. Russia also plans to launch several micro satellites.

Cubesats and Mini Satellites

Several organizations worldwide have taken up the challenge of making Cubesats, which are slightly bigger than a matchbox. Students from several universities have taken initiatives in this regard with promising results.

14

Space Observations

THE earth's atmosphere is opaque to a large part of the electromagnetic spectrum. This creates problems with ground-based observation of the cosmos. As a result, radiation in infrared, extreme ultraviolet, X-ray and gamma ray wevelengths cannot reach ground-based detectors. So, it is almost impossible to observe the sky in these wavelengths from the ground. The Space Age began in the 1950s, which opened up enormous possibilities of observing the cosmos in wavelengths that are inaccessible from the ground.

X-ray Stars

In 1895, German physicist Conrad Wilhelm Roentgen discovered the X-rays. Thereafter techniques have been developed for converting X-rays into visual images or into electronic signals that can be recorded. In medical science, X-rays are used only as a diagnostic tool, for imaging internal organs of the human body. X-rays are produced artificially by an X-ray machine, employing extremely high-voltage electricity. So, nobody expects that X-rays can also come from space. But, it is now a fact. The detection of cosmic X-ray sources is the evidence of the kind of energetic processes going on in the far-reaches of the cosmos. This is another example of space technology, which is changing our ideas about the universe we live in.

The first X-ray pictures of the sky were taken in 1947, just after the end of the Second World War when the Americans launched a captured German X-ray rocket to a height of 160 km. The rocket carried an X-ray camera, which took "pictures" of the sun. The first pictures were not spectacular, but they did show that the sun does emit some X-rays. The X-ray picture of the sun showed the regions in the sun's very thin outermost layer called "corona", which are extremely hot and actually emit X-rays. Later observations from the American orbiting space station Skylab showed the X-ray sources to be associated with other high-activity areas on the sun, such as sunspots and solar flares.

Apart from the sun, other sources of X-rays were subsequently discovered in space. Extremely hot gases in galaxies and star clusters are also known to give off X-rays, as do remnants of supernova explosions. Another copious source of cosmic X-rays is binary stars in which one of the components is a white dwarf, a neutron star, or even a black hole. All these sources emit X-rays by extreme acceleration of electrons – a process known as "synchrotron" radiation. In fact, emission of synchrotron X-rays offers an easy way of detecting black holes, which are themselves invisible, as no light can come out of them.

The discovery of an X-ray source outside the solar system came in 1962, when an American Ranger spacecraft was on a mission to the moon. It happend by chance to discover a very strong X-ray source in deep space. As it was discovered in the direction of the constellation Scorpius, the source came to be known as Sco X-1. The object was later found to be associated with a faint star and was emitting 1,000 times more energy as X-rays than at visible wavelengths. The discovery pointed to the fact that, besides our sun, there were certainly distant and more powerful X-ray sources as well.

The first satellite devoted to X-ray astronomy was launched in December 1970. Named Uhuru, the satellite detected more than 300 sources of X-ray emission in the sky during its two-year lifespan.

One major scientific result of Uhuru was the discovery that a large fraction of the cosmic X-ray sources detected is of binary systems, in which matter from a normal star is falling onto an extremely dense object, such as a white dwarf, a neutron star, or a black hole. These binary systems were found to emit almost 1,000 times more energy in X-ray wavelengths alone than the sun radiates in all the wavelengths. Later, as more refined techniques were developed and more advanced satellites were put into the orbit, more celestial X-ray sources were discovered. Prominent among them are the Crab Nebula in the constellation of Taurus and Cyg X-1 in the constellation of Cygnus.

Chandra in Orbit

In 1999, the US National Aeronautics and Space Agency (NASA) established the most powerful space-borne observatory to study X-rays from the stars. It is named Chandra X-ray Observatory after the Indian-born Nobel laureate Subrahmanyan Chandrasekhar (1910-95), who is known for his pioneering work on the evolution of stars. Its designers say, Chandra would unearth about 1,000 new X-ray sources in every patch of the sky.

Chandra X-ray Observatory contains a big telescope. The length of Chandra is 13.8 m. But what make Chandra really an advanced instrument are its mirrors that are used to focus X-rays. There are four sets of concentric barrel-shaped mirrors, which can bring X-rays from distant sources to a much sharper focus than ever possible. In other words, images received from Chandra will be much more detailed than any obtained so far. Sharper images would allow astronomers to see not only fainter objects but also finer details.

Another unique feature of Chandra is its orbit. Unlike all other satellites, including the Hubble Space Telescope, which is placed in a circular orbit, Chandra has been placed in a highly elongated orbit, which brings it within 10,000 km of the earth at its nearest approach and takes it as far away as 140,000 km at its farthest point in orbit every 64.2 hours. At its farthest point Chandra will be fully one-third of the way to the moon!

In the field of X-ray astronomy, Chandra represents a new beginning. The presence of a black hole at the centre of our Milky Way galaxy was already revealed, ushering in the era of routine high resolution X-ray imaging, and more important, of X-ray astrophysics. For the first time, it is allowing astrophysicists to apply the tools of physics to high energy astronomy. There is no doubt that Chandra's improved sensitivity will make possible more detailed studies of supernovas, black holes, and dark matter, and turn a new leaf in our understanding of the origin, evolution, and destiny of the universe.

A Quark Star

In April 2002, NASA announced the discovery of an entirely new king of stars unknown before. The announcement, based on results from the Chandra X-ray Observatory, said the new star is possibly entirely made of fundamental particles called "quarks". The unique feature about this star is that it is only a few kilometres across but weighs more than our sun. The star, called RX J-1856 is about 360 light-years from the earth and if it were really made of quarks, it would be the first example of its kind.

The Infrared Sky

In 1800, a famous English astronomer William Herschel (1738–1822) discovered the infrared radiation. It is a ubiquitous form of invisible electromagnetic radiation that is present almost everywhere. In fact, all bodies above the temperature of zero degrees absolute (minus 273.16° C) give off infrared radiation. We can feel infrared radiation as heat when it falls on our skin.

The study of the sky at infrared wavelengths could reveal the location and character of cool material in the cosmos – material that is not hot enough to give off light – such as clouds of gas and dust in the process of forming stars or planets, or a star nearing the end of its life.

On earth, infrared radiation is commonly associated with "warm" objects, but celestial objects that emit in these wavelengths are actually a great cooler than objects that emit visible light. For example, very young stars, which may not be visible through optical telescopes, are very bright infrared sources. Besides, the longer wavelength allows infrared radiation to penetrate cosmic dust found around or between stars and galaxies, which are opaque to much shorter visible wavelengths, thus allowing astronomers to look behind impenetrable clouds of cosmic dust. The wevelength of infrared radiation ranges from a few micrometres to a few hundred micrometres. (One micrometre is one millionth of a metre.) In fact, infrared radiation is longer than red light but shorter than microwaves. But, since the earth's atmosphere cuts off infrared wavelengths longer than a few micrometres, it is impossible to carry out infrared observation of the sky from the ground.

Till the advent of satellites, astronomers had used numerous techniques to go beyond the earth's atmosphere in order to have a better view of the sky in infrared wavelengths. Aircraft, balloons, and rockets were all used, but with only limited success. The first satellite exclusively devoted to infrared imaging of the sky was launched by NASA on 25 January 1983, called Infrared Astronomical Satellite (IRAS). It carried a 57-cm telescope cooled to a temperature of 2.5° above absolute zero and was operated jointly by the United States, the United Kingdom and the Netherlands.

During its six-month sojourn in space, IRAS sent back a spectacular, new view of the cosmos that revolutionized our understanding of the solar system, nearby stars, the Milky Way, and distant galaxies. Only a small part of the IRAS data has been analysed so far. It may take several years to process the entire data gathered by IRAS.

Ultraviolet Radiation

In 1803, the German physicist Johann Ritter (1776–1810) discovered ultraviolet radiation. He found that photosensitive crystals of silver iodide showed the maximum darkening when placed beyond the violet end of the spectrum. Our sun releases copious amounts of ultraviolet radiation, but most of it is absorbed in the upper atmosphere by the earth's ozone layer. Astronomers had known that high-energy processes, especially very high tempreatures, in cosmic bodies also release copious amounts of ultraviolet radiation. So, studies in ultraviolet wavelengths could provide new insights into cosmic processes. But, in this case too, the obstacle was the earth's atmosphere.

The first successful attempt to photograph the sun in ultraviolet was made in 1946, when a rocket-borne camera did the job. Since the early 1960s, the United States and several other countries have placed in the earth orbit unmanned satellite observatories for ultraviolet imaging of the sky.

In January 1978, NASA launched the International Ultraviolet Explorer Satellite, which opened up many facts about the ultraviolet radiation. As a joint project of NASA, the United Kingdom, and the European Space Agency, the satellite sent back data that supported the theory that a black hole with the mass of a thousand solar systems exists at the centre of our Milky Way galaxy.

The data also provided evidence of gravitational lensing by a massive galaxy bends the light coming from the distant quasar, as predicted by Einstein's relativity theory, to produce the double image. The IUE also discovered that our Milky Way galaxy is surrounded by a halo of hot gases.

Gamma Rays from Space

Gamma rays were named by the British physicist Ernest Rutherford (1871–1937) in 1903. They were the most energetic among the three kinds of radiation given off by radioactive elements. In space, they are produced by violent processes taking place in the interior of galaxies. For example, gamma rays are produced when lighter elements are created by nuclear fusion in the interior of stars and in supernova explosions, where heavier elements are created. Gamma rays are also produced in processes involving the interaction of matter and anti-matter and of subatomic particles with other particles or with magnetic fields, matter or even photons. Many of these processes are believed to occur in the extreme conditions found in and around supernovas, black holes, pulsars, and quasars. Although gamma rays are a highly penetrating form of radiation, they are readily absorbed by the earth's thick atmosphere. So, ground-based observation is almost impossible and observations have to be made from space.

Interestingly, the detection of gamma rays from space came about almost accidentally. In the late 1970s the United States had been launching earth satellites to detect gamma-ray emissions from nuclear testing. Surprisingly, these satellites instead discovered bursts of gamma rays from deep space. These occurred at a rate of one or two per day and typically lasted for a few seconds to a few minutes. But their source remained a mystery. Since then, several other spacecraft have been launched to probe gamma rays coming from space. Among them was the Compton Gamma Ray Observatory (CGRO) launched in April 1991. It carried instruments that are ten times more sensitive to gamma rays than earlier gamma ray satellites on board. By 1998, CGRO had detected more than 1,700 gamma ray bursts, distributed in a completely random pattern across the sky. Due to extremely short duration of the bursts, CGRO was unable to locate their position accurately.

In 1990, an Italian–Dutch satellite, called BeppoSAX, was launched into the orbit. It was able to achieve a fiftyfold increase in positional accuracy; that is, it could locate the source of the gamma ray bursts much more acccurately than CGRO could. In December 1997, BeppoSAX detected the most powerful gamma ray burst till date, coming from a source located some 12,000 million light-years away. The powerful burst is believed to have come from the merger of two neutron stars. To explore and understand the cosmos, the data sent back by CGRO and BeppoSAX helped a lot.

15

Radio Waves from Space

As we know today, visible light is a form of electromagnetic radiation and makes up only a very small part of the electromagnetic spectrum. The human eye is sensitive to only this limited range of wavelengths — corresponding to the colours red to violet. This is why our eyes cannot perceive radiation outside this range, which comprise the whole range of wavelengths from infrared to radio waves beyond red, and from ultraviolet to gamma rays beyond violet. The wavelengths of visible light and other electromagnetic spectrums are different. Radio waves have relatively longer wavelengths, while gamma rays have much shorter wavelengths compared to visible light. But, radio waves can be received by using special antennas. In fact, all radio receivers work by receiving radio waves using an antenna and then separating the sound signal and amplifying it by employing an electronic circuit. The signals that we receive in our radio set are sent out by transmitters on earth. Till the early 1930s, nobody could imagine that radio signals could also come from outer space.

Radio Waves from Space

In December 1931, the first source of radio waves from space was discovered, during a routine study of static interference in radio transmissions, a common phenomenon, which adversely affects the quality of the signals received. In the 1920s, amateur (ham) radio operators across the world were establishing communication links among themselves by using short-wave frequencies. But, such wireless telephone links were found to be plagued by atmosphere-static noise caused by lightning and other atmospheric disturbances, which badly affected the transmission quality.

During that time, a young radio engineer named Karl Jansky (1905-50), working with AT&T Bell Labs in New Jersey, USA, searched for the cause of the short-wave noise. He built a highly directional antenna to work at about 22 MHz (i.e. capable of receiving radio waves having a frequencey of 22 million cycles per second) and began to make systematic observations. Most of the noise he found was due to thunderstorms and other terrestrial causes. Jansky had found that the radio noise was emitted from the centre of our own Milky Way galaxy. He made this discovery in 1932 and announced his findings in 1933. However, Jansky's discovery of radio waves coming from space did not attract the attention of astronomers immediately. But, there was one individual who took notice of it. He was Grote Reber (1911–2002), another radio engineer, who was an avid ham operator. Reber had spent much time making long-distance contacts on the amateur short-wave bands. He had "worked" all continents and sixty foreign countries. In those days, that was quite an achievement, and it left Reber thinking, as he later wrote, "there were no more worlds to conquer". When he read of Jansky's discovery, he realized that he had still a long way to go.

In 1937, Reber decided to take up Jansky's work as an opportunity and challenge. In his spare time, with his own resources, he built for himself a steerable parabolic dish antenna (like the ones cable operators use to receive TV signals from satellites) in his backyard to scan the sky for radio signals. Reber's antenna was quite large, almost 10 m in diameter. In an era when nobody had even dreamt of artificial satellites and television had not yet emerged from the laboratory, this antenna became a public curiosity, drawing amazed remarks from his neighbours.

Reber continued his observations of the radio sky for more than a decade. His observations produced an amazing view of the universe. His radio "maps" showed for the first time the startling differences between the "visible" sky and the "radio" sky. It marked the dawn of an entirely new branch of astronomy, which came to be known as "radio astronomy".

Radio Telescopes

Radio astronomy, or the study of the cosmos in radio wavelengths, soon became an established subject and radio telescopes, with large dish antennas, were discovering new phenomena in the universe. The first really large, fully steerable, radio telescope was completed in 1957 at Jodrell Bank, England. Further developments in the radio telescope were done by the different organizations worldover.

Since radio waves are much longer than light waves, the resolution of radio telescopes is rather limited. Therefore, radio astronomers use an innovative method, known as "radio interferometry" to study the finer details of a radio source.

Several celestial objects like the planets and the stars emit the radio waves. By 1950, several other sources of radio waves were identified. They included the famous Crab Nebula, and two galaxies – M87 and NGC5218. Later, radio telescopes enabled planetary scientists to discover intense radio emissions from Jupiter and to measure the temperature of all the planets. By the mid-1980s some 100,000 cosmic radio sources had been catalogued.

Utilizing radio telescopes, equipped with sensitive microwave spectrometers, researchers have discovered more than fifty separate moleclues, including familiar chemical compounds like water vapour, formaldehyde, ammonia, methanol, ethanol, and carbon dioxide in space.

Powerful "Star-like" Sources

Amazing discovery in radio astronomy came in 1960, when an entirely new class of astronomical objects never known before was detected in deep space. Out of the first radio "stars" discovered in 1960, that could be identified with a star in a photographic image, was called "3C48". It turned out to be an entirely new class of astronomical objects known as "quasi-stellar radio sources" or quasars. These are point-like objects which appear like stars in visible light, but also emit extremely strong radio waves. A typical quasar was found to be no more than a light-year or two in size, but up to 1,000 times more luminous than giant galaxies with a diameter of about 100,000 light-years. Measurement of the red shift of the light from these objects showed them to be extremely distant, a typical distance being more than 10,000 million light-years.

By the end of the 1960s, radio atronomers had discovered more than 150 quasars, although not all of them were found to emit strongly in radio wavelengths. The nearsest quasar, at a distance of only 783 light-years, was discovered in 1975. Then the astrophysicists found the rapid pulsation of the radio signals from some strange cosmic objects, which were called pulsars.

Relics from the Past

Quasars and pulsars were not the only objects discovered after the advent of radio astronomy; it also provided a much-awaited proof of the Big Bang — the colossal event that is supposed to have brought this universe into existence in a primeval fireball, some 10,000 to 20,000 million years ago. In that titanic cosmic explosion, the universe began to expand, which has never ceased. As early as 1948, the famous American physicist George Gamow (1904-68), in a paper he wrote with Ralph Alpher (1921–2007), made a remarkable prediction that, if, indeed, there were a Big Bang, the radiation accompanying it must have cooled down with the expansion of the universe and should still be around in the form of microwave radiation. It was predicted that the microwave background radiation should be characteristic of objects at a temperature of about 5 K (i.e. 5° above absolute zero) and should be coming from all parts of the sky as a homogeneous background.

The elusive, all-pervading background radiation was eventually detected by two scientists of the Bell Telephone Laboratories, viz. Arno Penzias (1933–2024) and Robert W. Wilson (1936-) in May 1964. Here too, the actual discovery occurred by accident. While conducting experiments with the first Telstar communication satellite, Penzias and Wilson detected excess of radio noise that seemed to come uniformly from all parts of the sky. The measured temperature of the radiation turned out to be 3 K.

In 1989, a satellite called Cosmic Background Explorer (COBE) was launched to make detailed measurements of cosmic microwave background radiation. The COBE data provided the first evidence of condensation of galaxies in early universe.

Large Arrays of Antennas

Radio waves are more than 10,000 million times longer than light waves. So, to have the required resolving power, a radio telescope ought to have a dish antenna 10,000 million times bigger. But, this is an absurd idea. So, the radio astronomers decided to use a large array of smaller dishes, which together could act as a single antenna of enormous size. The largest array in operation today is the Very Large Array (VLA) radio telescope set up in New Mexico in the US.

In 1980, VLA was operated. It was made with twenty-seven parabolic dishes, which are each 25 m in diameter. Each of these dishes is mounted on a transporter that can be moved along rails laid down to an enormous Y pattern. Each arm of this pattern is about 21 km long. The radio signals recorded by the component dishes are integrated by a computer, so that the entire array acts as a single radio antenna with a maximum effective aperture of 27 km, that is, the array behaves like a single dish antenna, 27 km in diameter, which would otherwise be impossible to build. This large aperture gives VLA a resolving power equal to that of the best ground-based optical telescopes.

The world's largest radio telescope array is in India (at Narayanagaon Pune). Known as the Giant Metrewave Radio

Telescope (GMRT), it is an array of thirty fully steerable parabolic dish antennas, each 45 m in diameter. Twelve of these are located in a compact central array, about 1 km in size. The remaining eighteen antennas form the three arms of an approximately Y-shaped configuration, with each arm extending to about 14 km from the array centre. Fourteen of the thirty antennas had been fully commissioned till the end of 2001: eleven in the central array and three in the Y arms of the arrray. Giant Metrewave Radio Telescope (GMRT) is now under research for development. It, when completed, will become the world's most powerful radio telescope operating in the frequency range of about 50 to 1500 MHz. The large size of the parabolic dishes implies that GMRT will have over three times the collecting area of VLA. GMRT will be about eight times more sensitive than VLA, because of the larger collecting area, higher efficiency of the antennas, and a substantially wider usable bandwidth. Already some startling observations of supernova remnants and radio galaxies have been made, using GMRT.

Indeed, radio telescopes like VLA and GMRT are revolutionizing our ideas about the cosmos. They are also throwing new light on the origin of the universe. In the latter half of the twentieth century, advances in instrumentation and improved observational techniques, especially space-borne detectors, have led to the discovery of cosmic sources that emit infrared, X-ray and gamma ray wavelengths. Normally, these wavelengths do not reach the ground as they are cut off by the earth's atmosphere. Discovery of radio telescopes has radically changed our view of the universe and of the cosmic phenomena going on there.

16

Role of Physics in Information and Communication Technology

Prelude

PHYSICS is that branch of science which deals with matter and energy in space and time. It owes its origin to the significant event Big Bang and the creation of the material world. In course of time, physics is developed with new discoveries and inventions. It is said that where physics is applied that becomes "technology". In the present age of science and technology, we come across new invention or innovation only by the use of physics. For the comfort of our daily life, we use a range of products like bulbs, fans, coolers, air conditioners, washing machines, televisions, fridges, mixers, printing machines and microwave ovens which are running on electricity, a historical invention of physics. The principles of physics and its ability to analyse the phenomenon is the backbone of current and future developments. But, now physics is applied to new developments and innovations in technology which led to improved quality of human life. The discovery of the diagnostic tools like X-rays, Magnetic Resonance Imaging (MRIs), surgical equipments, cancer treatment with radio isotopes, endoscopy with optical fibres is all based on physics. Information and

Communication Technology (ICT) is the latest gift of this discipline. The invention of transistors, videos and digital cameras, computers, laptops, rocket launching in space, drones, Light Emitting Diode (LEDs), smart phones, smart cars, and smart air planes are some examples of advanced technology based on the principles of physics.

Involvement of Physics in ICT

The main components of ICT are people, data, hardwares, software, information and procedure. People are the driving force of this system and involved at various stages of ICT network. They are needed to collcet the non-digital data in the form of text, numbers, images, photographs, graphs, audio and video recording, etc. from various sources and enter into the ICT system through keyboard/ scanner or by uploading audio/video recordings, thereby involved in collecting the relevant information, as per their requirements. People who enter the data and use the information are called end-users. The other main users are database administrators, database designers, system administrators and network administrators.

ICT revolution has brought a drastic change in our everyday life. Transistors, computers, and Internet are considered to be the most significant invention of the twentieth century and lay the foundation of modern technology. Today, ICT proved its mettle at social, economic, political and business platforms, and is connecting the entire economy of the world; but, the core of these technologies is physics which gives the basic understanding behind the various advancements. The inventions in different fields of physics are further carried with the development of modern technologies.

Different branches of physics that build ICT are the following:

1. Vacuum tubes,
2. transistors,
3. semiconductor physics,
4. integrated circuits,
5. very Large Scale Integration (VLSI),
6. electromagnetic waves,

7. laser,
8. optical fibre,
9. ultrasonic waves,
10. nanotechnology,
11. nano-electronic devices,
12. opto-electronic devices,
13. quantum computers,
14. magnetic materials and
15. superconductivity.

Epilogue

Physics plays a crucial role in the economic development of any country and ICT proves to be the strongest pillar of this development. Through various branches of physics, ICT grows and by that the nations progress. Today's century is fast, modern, and advanced, where physics is magnificently performing its role from solid state and material science to nanotechnology and superconductivity. Scientists wonder that where would these advancements end up as it is against the law of nature. But these are exceptional technology efforts. Scientists know the limit of physics and we are near to achieve this limit in this revolution. We hope that the future will witness the exploration of more ways of communication.

17

Genesis and Growth of Meteorology

In 350 BCE, Aristotle, the famous Greek scholar, formed the earliest systematic treatise on meteorology. Even after several centuries, no correct predictions could be made on climate and weather. Then Galileo Galilei invented the first instrument useful for weather study, the thermometer. His student Evangelista Torricelli followed it up by a barometer in 1643. It was only in the eighteenth century that the nature of trade winds and the effects of the earth's rotation on them were understood. In 1648, Blaise Pascal (1623-62) noted that the air becomes less dense with height. In 1714, a German physicist, Daniel Gabriel Fahrenheit (1686–1736), used mercury in a thermometer. Thirty years later, Andres Celsius (1701-41), a Swedish astronomer, announced the centigrade scale for temperature measurement.

Joseph Black (1728-99) formed carbon dioxide in 1754. This was followed by the discovery of hydrogen in 1766 by Henry Cavendish (1731–1810) and nitrogen in 1772 by Daniel Rutherford (1749–1819). In 1774, Joseph Priestley (1733–1804) found oxygen, the name given by Anteine Lavoisier (1743-94). Also, Wilhelm Sheele (1742-86), who incidentally discovered oxygen two years before Priestley isolated the gas chlorine (though it was not published).

Even when the Montgolfier brothers [Joseph Michel (1740–1810) and Jacques Étienne (1745-99)] in France rode a balloon over

Paris in 1783 and became the world's first aeronauts, there was little realization that the air was like a "fluid" which allowed objects to float. Nevertheless, balloons for weather observation were sent up. In the middle of the nineteenth century, the loss suffered by the French navy in a storm during the Crimean War (1853), prompted a French astronomer, Joseph Leverrier (1811-77), to damonstrate that the path of storms could be traced and even predicted by means of daily weather reports from different stations. His optimism was sparked by the telegraph invented only a decade earlier (1844) by Samuel Morse.

Several institutions were established towards the end of the nineteenth century to study the weather. In 1875, the Indian Meteorological Department was established. France used weather balloons to study the upper air. In the first two decades of nineteenth century, new approaches to predict weather, based on mathematics and statistics, were attempted. After the First World War, aeroplanes were used for weather observations in the US, where teletypewriter weather circuits were also established. The Second World War saw the introduction of radar which was used later for storm detection. Around 1946, some cloud-seeding experiments were made in the US, followed by the use of sounding rockets for weather observations in 1950. With the invention of the computer, experiments were conducted using it for weather study at the Institute of Advanced Study in Princeton (US) and by many others. The Space Age gave new ideas and prompted scientists to send their instruments into earth orbit.

Satellite meteorology began with the American Television Infrared Observation Satellite (TIROS-1) launched in April 1960 in near-earth orbits. The satellites in this series looked away from the earth most of the time and did not pass near the poles. The method of transmitting cloud pictures was improved in 1963 when simple low-cost ground receivers were able to get them in a matter of about 3 minutes. The method, known as Automatic Picture Transmission (APT), soon became popular in many countries. Satellite imagery was freely available in real time anywhere in the world to anyone who could have the receiver. As a result of advances in technology, infrared sensors were tried to detect radiation from clouds and land. Near-polar orbiting, sunsynchronous satellites were developed to carry the sensors. The second-generation weather satellites could take cloud pictures both during day and night. The TIROS series (1960-65) and the nine Environmental Science Services Administration (ESSA) satellites (1966-69) were followed by improved TIROS (1970). The experimental series was succeeded in 1979 by the US National Oceanic and Atmospheric Administration (NOAA) satellites. Two NOAA satellites operate in complementary near-polor sunsynchronous orbits at an altitude of 854 km. NOAA-18 is operational since August 2005 with two back-up satellites. NOAA data now make it possible to predict the weather related to El-Nino for a season or more than a year in advance. China and Russia also have polar-orbiting weather satellites.

In order to develop the weather monitoring systems, many other advanced sensors and satellite systems have been created. Among them the Nimbus series started by the US, Astro Vision of Australia, and METEOSAT of Europe are notable. The world Meteorological Organisation was established in 1963. The World Weather Watch, aimed primarily at detecting major atmospheric phenomena and dissemination of data was started in 1963.

18

NASA's Five Significant Missions

Prelude

THIS universe of ours is composed of mainly five basic elements, viz. earth, air, fire, water, and *antarikṣa* (space). Man, by nature, is curious to know about the space, its nature, and characteristics. Man observes the events happening in nature, like the shining of the sun, the beaming of the moon, the twinkling of the stars, rising and setting of the sun along the horizon, changing the shape of the moon, the lightning in the sky, formation of clouds, roaring of thunder and raining, creation of tides in the sea, and flying of birds in the sky. The events happening in the space attract the human mind to know more about the space beyond the earth. First of all, man wanted to fly like a bird. He prepared gliders and tried lift himself to the space. This dream of man came true when the Wright brothers invented the flying machine and finally aeroplane came to exist. Gradually, with the advancement of technical skill, man prepared satellites and launched them into space for setting up communication among different nations beyond the ocean barrier and distance hurdles. He used the satellites for weather forecasting and monitoring.

There exist also some natural satellites of the planets. The moon is the natural satellite of the planet earth. Man became interested to know about the planets beyond the earth and their satellites. Man became curious to explore the space beyond the earth. He launched space expeditions through unmanned probes. In course of time, with the advancement of technology, astronauts were sent to the space. The Soviets won the race in April 1961, when cosmonaut Yuri A. Gagarin completed a single orbit around the earth aboard his Vostok capsule. On 5 May 1961, Alan B. Shepard became the first American in space during a suborbital flight aboard in his Mercury capsule named Freedom-7. After achieving success in sending man into space, space scientists tried to launch missions to space for investigating the physical phenomena happening in the moon, Mars, Venus, Jupiter, Saturn, etc.

Historic space missions include Project Gemini, Apollo-11, Viking-1, Apollo, Hubble Space Telescope, Apollo-8, Apollo-13, Voyager-1, Skylab, Sputnik-1, Messenger, Mars Odyssey, Mars Reconnaissance Orbiter, Dawn, Pioneer-10, Apollo-15, Apollo-7, Voyager-2, Galiteo, Magellan, Apollo-Soyuz, Pioneer-11, and Apollo-14.

Some most important space missions in history are:

1. 1957: Sputnik-1 – first satellite in space.
2. 1961: Vostok-1 – first man in space.
3. 1963: Vostok-1 – first woman in space.
4. 1965: Mariner-4 – first image of another planet.
5. 1969: Apollo-11 – first man on the moon.
6. 1970: Voyager-7 – first landing on another planet.
7. 1970: Apollo-13 – a daring rescue.
8. 1977: Voyager-1 – mission to interstellar space
9. 1986: STS-51-l – the challenger disaster.
10. 1990: Hubble Space Telescope – spying on the stars.
11. 1996: Mars Pathfinder – first rover on mars.
12. 1997: Cassini-Huygens – sailing around Saturn

13. 2000: The International Space Station – a critical laboratory.
14. 2009: Kepler – exploring distant worlds.
15. 2015: Falcon-9 – lands safely.
16. 2016: The Juno Spacecraft, which successfully entered the orbit of Jupiter on 4 July 2016, will for the first time peer below the dense cover of clouds to answer questions about the gas giant, origins of our solar system.
17. 10 April 2019: First direct photographs of a black hole and its vicinity by Event Horizon Telescope.
18. 19 April 2021: First aerodynamically powered flight on another celestial body (Mars) by ingenuity.

Significant Space Missions of NASA

The National Aeronautics and Space Administration (NASA) of USA has created a long record of successful space programmes, which have unveiled our many doubts and questions on our solar system — and the universe itself. Such continuous efforts of NASA go a long way for our sicentists as well as the general public. The achievements in space exploration help us expand our knowledge about the space beyond the earth and the celestial bodies as well.

From achieving what was thought impossible, to showering us with awe-inspiring images of distant worlds, NASA's scientific endeavours have filled us with a vast array of emotions from booming pride, to feeling very special and utterly insignificant.

Let us focus on the five most remarkable undertakings of NASA in chronological order.

Apollo

America's technological advancement is shown by NASA's Apollo mission. The Apollo programme would become NASA's greatest and most famous mission, achieving the impossible by landing a man on the moon.

Named after the most loved of all the Greek gods Apollo, the Apollo programme consisted of seventeen missions which continued from 1961 to 1972.

Apollo-1 ended in a major tragedy when a fire broke out during a simulation phase and three astronauts lost their lives. Apollo mission 2-6 were unmanned test flight, while Apollo-7 became the first crewed mission of the programme designed to test command and lunar modules.

Apollo-8 was the first manned spacecraft to leave low-earth orbit and to orbit the moon. Landing on it became just a matter of time.

Apollo-9 and Apollo-10 were again crewed test flights, but Apollo-11 became the most significant one, the first to land a human on an extraterrestrial body.

Apollo-11 was launched from Cape Conaveral on 16 July 1969 and touched down on our only natural satellite four days later, Neil Armstrong and Edwin Aldrin walked on the moon, while Michael Collins stayed in the command module. The astronauts took photographs, collected samples, and returned safely back home on 24 July. They left behind the plaque that reads:

> Here men from the planet earth first set foot on the moon July 1969 A.D. We came in peace for all mankind.

Five of the subsequent six Apollo missions were successful in landing a crew on the moon, except for Apollo-13 which failed the landing but still managed to return home safely after an explosion on the ship.

The crew of Apollo-15 and Apollo-16 crew were the first to drive a car on the moon, while the last mission — Apollo-17 — was the longest one. The astronauts stayed on the moon for more than three days and performed these extensive moonwalks.

In three short years. the Apollo programme landed twelve men on the moon and remains the only mission to have achieved this.

Humanity has never returned to the moon afterwards, nor has it managed to land on any other object beyond our beloved planet.

Pioneer-10 and Pioneer-11

Pioneer-10 and Pioneer-11, launched in 1972 and 1973 respectively, were the first spacecraft to visit the solar system's most photogenic gas giants, Jupiter and Saturn. Pioneer was the first probe to travel through the solar system's asteroid belt, a field of orbiting rocks between Mars and Jupiter. Then about a year and a half after its launch, the spacecraft made the first flyby of the planet Jupiter. It took stunning very close photos of the great red spot and the wide swaths of red that band the planet. About a year later, Pioneer-11 flew by Jupiter and then moved on to Saturn, where it discovered a couple of previously unknown small moons around the planet and a new ring. Both probes have stopped sending data and are continuing out on their one way voyages beyond the solar system.

Launched in 1972, Pioneer-10 was the first NASA mission to the outer planets of the solar system. It was the first spacecraft placed on a trajectory to leave the solar system and venture into the interstellar space. It was also the first space probe to fly beyond Mars, through the main asteroid belt, and to cross the orbits of Jupiter and Neptune.

Pioneer-10 misson was initially planned for a twenty-one month duration for a flyby of Jupiter, but it lasted more than thirty years. It sent its last signal to earth in January 2003 when it was almost 8 billion miles (12 billion km) away.

Pioneer-11 was launched after one year of the previous mission. It was the first space probe to fly-past Saturn. The last contact with the probe was established in September 1995.

The Pioneer crafts look numerous images of Jupiter and Saturn, which helped to uncover many mysteries surrounding the gas giants, their composition and their moons.

Pioneer-10 and Pioneer-11 both carry small aluminium plaques with our location in the galaxy and a drawing of a naked man and a woman. The brain child of Carl Sagan, these plaques would provide any alien civilization that found them in a distant future with the basic information on the civilization that created us.

Today, these space probes are no longer in communication with earth, but they are still moving farther and farther away from us. Pioneer-10 is travelling towards the Taurus constellation, while Pioneer-11 is heading towards the constellation of Aquila. It will take both of them several million years before they encounter first stars.

Voyager-1 and Voyager-2

Voyager-1 and Voyager-2 were launched just a few years after the Pioneers. These missions were aimed at studying the gas gaints of our outer solar system.

Voyager-1 was sent off after Voyager-2, but overtook its twin ship in December 1977 and reached the Jovian system in April, the following year. It took stunning photos of Jupiter and its two moons, discovering two new ones. The biggest surprise was noticed when images of the moon Lo revealed a strange yellow-orange and very geologically active celestial body.

The probe reached Saturn after two years and discovered five new satellites and showed that most were made largely of ice. It also provided significant data on Saturn's biggest moon Titan, which the Cassini–Huygens mission would explore a few decades later. After this encounter with Saturn, Voyager-1 started a trajectory to escape the solar system at a speed more than 500 million km (300 million miles) per year. In 1998, it became the most distant human-made object when, at a distance of 70 A.V. from the sun, it overtook Pioneer-10.

In 1990, while heading towards the distant Cosmos, Voyager-1 turned around and took one last photograph of the solar system, which would be known as "The pale Blue Dot".

Voyager-2 is the only spacecraft designed to study all four gas giants, viz. Jupiter, Saturn, Neptune, and Uranus. It discovered many new moons lurking around these planets. It sent us spectacular images of the Jovian and the Saturnian systems, and detailed photographs of Saturn's rings. In 1985, it became the first human-made object to fly past Uranus and in 1989, the first to reach Neptune. It relayed back many new facts about these gaseous planets and their satellites, as well as astounding imagery.

Both Voyager spacecrafts carry a photograph record called the Golden Record. Like the Pioneer's plaques, these discs contain information about their origins, in the form of images, natural sounds, music, and greetings in fifty-five languages.

The Voyagers continue to transmit data to earth every day. When eventually they run out of electricity to power their instruments, they will silently continue their eternal journey through the cosmos.

Hubble Space Telescope

Put into earth's lower orbit in 1990, Hubble revolutionized the way general public envisions the cosmos' breathtaking images inspire us and make us appreciate the glory and vastness of the universe we live in.

We are astonished at the amazing feats of creation by observing the Eagle Nebula or the Hubble Deep Field showing thousands of galaxies. We are thinking of all the staggering worlds and potential civilizations that might thrive there, as unaware of our existence here in this universe.

Placed at 500 km (350 miles) above the earth's surface, Hubble is far above the atmospheric distortions that blur the vision of ground telescopes. It has an unobstructed view of our solar system, distant galaxies as well as the far edge of the universe.

Named after American astronomer Edwin Hubble, the Hubble Space Telescope is a true-time machine. We see distant objects as they were millions of years ago, as this is how long it took for light to travel from them and reach us. The Andromeda galaxy being 2.5 million light-years from the earth, thus appears as it was 2.5 million years ago.

By measuring a special kind of stars known as Cepheid variables, Hubble has helped astronomers refine the age of the universe, narrowing it down to 13.7 billion years. It also observes stars in various stages of their evolution and even enable us peek into our neighbours – Magellanic clouds and Andromeda galaxy.

This telescope has helped us demystify black holes, determine the rate of the universe's expansion, and create a 3-D map of dark matter. It also provides stunning photos of our closest neighbourhood – the planets in our solar system. It has recorded comets crashing into Jupiter, water erupting from Jupiter's moon Europa, and discovered two of Pluto's moons.

The Hubble Space Telescope is one of NASA's most exceptional missions and has been inspiring the curious minds for the past sereval decades.

Cassini–Huygens

It was launched in January 1997. The Cassini spacecraft's objective was to study Saturn, its rings, and moons. Travelling with a passenger – the European Space Agency's Huygens probe – it would explore the Saturnian system for twenty years.

Cassini reached Saturn in 2004 and released Huygens to Saturn's largest moon Titan just a few months later.

Named after the seventeenth-century Dutch astronomer who discovered Titan, the Huygens probe would become the f irst man-made object to land on an outer solar system body. Because of Titan's thick atmosphere, astonomers had no idea on what kind of surface the probe would be landing, making the challenge even greater.

Upon touch down, Huygens discovered an earth-like world of wet sand and water-ice pebbles. The probe was equipped with several instruments that helped cast light on the satellite's surface and atmosphere composition. It continued to send data back to earth for 72 minutes before losing contact with Cassini. This probe, on the other hand, would send data back to earth for another two decades. It discovered two new moons of Saturn, evidence of liquid water on one of the satellites — Enceladus, a giant methane lake on Titan — and numerous new facts about Saturn itself.

In 2016, the spacecraft was sent on a series of final flybys of Saturn to provide close-up views of the rings. On the ground finale day of 15 September 2017, it plunged into Saturn as predicted by the mission's goad sending as data until the very end.

Epilogue

The above-mentioned missions are a few NASA's most successful endeavours in their attempt to unveil the mysteries of the cosmos and to bring its wonders closer to us, but we could of course cite many more over thirty years, NASA's space shuttle fleet — Columbia, Challenger, Discovery, Atlantis, and Endeavour — flew 135 missions and carried 355 different people to space.

In 2020 space flight, total orbital launches of NASA is 114, out of which 104 were successful. First launch was on 7 January and last launch was on 29 December.

On 10 April 2019, first direct photograph of a black hole and its vicinity by Event Horizon Telescope were taken. In April 2021, first aerodynamically powered flight, on another celestial body (Mars) by Ingenuity was launched.

Probably, the most famous of all missions — Apollo — put the first men on the moon. The Pioneer and Voyager spacecraft gave us unprecedented data on the outer solar system and are travelling towards other stars, carrying information about their origins.

Cassini spacecraft and its probe Huygens helped us demystify Saturn and its moons, some of which might have conditions to host life. The Hubble Space Telescopes unequalled imagery has been filling us with amazement for decades.

Existing new missions, such as James Webb Space Telescope and new Mars rovers, are just around the corner. We are eagerly waiting to see what other wonders they will uncover.

Space view in 2022 is forwarding us to exciting space news, which is worth sharing with the sky watch enthusiastics. In 2022, an odd type of exploding star called a red nova would appear in our skies. This story really began ten years ago, when astronomers closely monitored a distant star in Scorpius.

19

Moon Missions
A Bird's-Eye View

Prolegomenon

THERE are so many moon missions launched by different countries in different times. Human beings are interested to know about our outer worlds, like stars, planets, satellites, asteroids, and comets. As a part of exploration of the moon, numerous space missions have been undertaken to study the earth's natural satellite. Of the moon landing, Luna 2 of the Soviet Union was the first spacecraft to reach moon's surface successfully, intentionally impacting the moon on 13 September 1959. In 1966, Luna-9 became the first spacecraft to achieve a controlled soft landing while Luna-10 became the first mission to enter the orbit.

Between 1968 and 1972, crewed missions to the moon were conducted by the United States as a part of the Apollo programme. Apollo-8 was the first crewed mission to enter the orbit in December 1968 and it was followed by Apollo-10 in May 1969. Six missions landed humans on the moon, beginning with Apollo-11 in July 1969, during which Neil Armstrong became the first person to walk on the moon. Apollo-13 was intended to land, however, it was restricted to a flyby due to malfunction aboard the spacecraft. All other nine crewed missions returned safely to the earth.

While the United States focused on the crewed Apollo programme, the Soviet Union conducted uncrewed missions that deployed rovers and returned with samples to the earth. Three rover missions were launched, of which two were successful and eleven sample return flights were attempted with three successes.

Missions to the moon have been conducted by the following nations and entities (in chronological order): The Soviet Union, the United States, Japan, the European Space Agency, China, India, Luxemburg, and Israel. The moon has also been visited by five spacecrafts not dedicated to studying it, four spacecrafts have flown past it to gain gravity assistance, and a radio telescope, Explorer-49 was placed, into the selenocentric orbit in order to use the moon to block interference from terrestrial radio sources.

Soviet Union's Moon Missions

The Soviet crewed lunar programmes were a series of attempts pursued to land humans on the moon, in competition with the United States' Apollo Programme to achieve the same goal set publicly by President John F. Kennedy on 25 May 1961. The Soviet government publicly denied participating in such a competition, but secretly pursued two programmes in the 1960s: crewed lunar flyby missions using Soyuz-7K-L1 (Zond) spacecraft, launched with the Proton-Krocket, and a crewed lunar landing using Soyuz-7K — LOK and LK spacecraft launched with the N1 rocket. Following the dual American successes of the first crewed lunar orbit on 24-25 December 1968 (Apollo-8) and the first moon landing on 20 July 1969 (Apollo-11) and a series of catastrophic N1 failures, both Soviet programmes were eventually brought to an end. The proton-based Zond programme was cancelled in 1970 and the N1-L3 programme was de facto terminated in 1974 and officially cancelled in 1976. Details of both Soviet progranmes were kept secret until 1990 when the government allowed them to be published under the policy of glasnost.

Despite the Soyuz-1 setback, the Soviets succesfully rehearsed, the automated docking of two uncrewed Soyuz craft in earth orbit in 1968 and with the crewed Soyuz-4 and Soyuz-5 joint mission in early 1969 tested the other key mission elements. A total of eighteen missions were related to the N1-L3 project. The Luna-9 spacecraft, launched by the Soviet Union, performed the first successful soft moon landing on 3 February 1966. It represented the first lunar sample return mission by the Soviet Union, and was the third lunar sample return mission overall, following the Apollo-11 and Apollo-12 missions of America.

A moon landing is the arrival of a spacecarft on the surface of the moon. This includes both crewed and robotic missions. The first human-made object to touch the moon was the Soviet Union's Luna-2 on 13 September 1959.

Luna Programme

The Luna programme (from the Russian word Лyha "Luna" meaning "Moon"), occasionally called Lunik by Western media, was a series of robotic spacecraft missions sent to the moon by the Soviet Union between 1959 and 1976. Fifteen were successful and twenty-nine flights met failures, each designed as either an orbiter or a lander, and accomplished many experiments, studying the moon's chemical composition, gravity, temperature, and radiation.

Soviet Union's uncrewed exploration of the moon was completed between 1958 and 1976. First flight was launched on 23 September 1958 by the vehicle Luna-E-1 No. 1. Last flight was launched on 22 August 1976 by Luna-24. The launch site was the Baikonur Cosmodrome. The launch vehicles were Luna-8K72 Molniya-L, Molniya-M, and Proton-K/D-1.

Twenty-four spacecrafts were formally given the Luna designation, although more were launched. Those failed to reach orbit were not publicly acknowledged at the time and not assigned a Luna number. Those that failed is low-earth orbit were usually given cosmos designations. The estimated cost of the Luna programme in 1964 was US $6-10 billion.

After the unsuccessful attempt by Luna-1 to land on the moon in 1959, the Soviet Union performed the first hard moon landing— "hard" meaning the spacecraft intentionally crashed into the moon — Later that same year with Luna-2 spacecraft, a feat the US duplicated in 1962 with Ranger-4, was impacted. Since then, twelve Soviet and US spacecraft have used braking rockets (retrorockets) to make soft landings and perform scientific operations on the lunar surface, between 1966 and 1976. In 1966, the USSR accomplished the first soft landing and took the first pictures from the lunar surface during the Luna-9 and Luna-13 missions.

The Soviet Union achieved first uncrewed lunar soil sample return with the Luna-16 probe on 24 September 1970. This was followed by Luna-20 and Luna-24 in 1972 and 1976 respectively. Following the failure at launch in 1969 of the first Lunokhod, Luna E-8 No. 201, the Luna-17, and Luna-21 were successfully uncrewed lunar rover missions in 1970 and 1973.

Many missions were failures at launch. In addition, several uncrewed landing missions achieved the lunar surface but were unsuccessful, including Luna-15, Luna-18, and Luna-23, all crashed on landing.

United States' Moon Missions

SURVEYOR PROGRAMME

The Surveyor programme was a NASA programme that, from June 1966 through January 1968, sent seven robotic spacecraft to the surface of the moon. Its primary goal was to demonstrate the feasibility of soft landings on the moon. The Surveyor Craft was the first American spacecraft to achieve soft landing on an extra-terrestrial body. The missions called for the craft to travel directly to the moon on an impact trajectory, a journey that lasted 63 to 65 hours, and ended with a deceleration of just over 3 minutes to a soft landing.

The programme was implemented by NASA's Jet Propulsion Laboratory (JPL) to prepare for the Apollo programme, and started in 1960, JPL selected Hughes Aircraft in 1961 to develop the spacecraft system. The total cost of the Surveyor programme officially was $469 million.

Five of the Surveyor spacecraft successfully soft landed on the moon, including the first one. The other two failed; Surveyor-2 crashed at high velocity after a failed mid-course correction and Surveyor-4 lost contact (possibly exploding) 2.5 minutes before its scheduled touch-down. All seven spacecraft are still on the moon; none of the missions included returning to earth. Some parts of Surveyor-3 were returned to earth by the crew of Apollo-12, which landed near it in 1969. The camera from this craft is on display at the National Air and Space Museum in Washington DC.

APOLLO PROGRAM

The Apollo program, also known as project Apollo, was the third US human space flight programme carried out by NASA, which succeeded in preparing and landing the first humans on the moon from 1968 to 1972. It was first conceived during Dwight D. Eisenhower's administration as a three-person spacecraft to follow the one-person project Mercury, which put the first Americans in space. Apollo was later dedicated to president John F. Kennedy's national goal for the 1960s of "landing a man on the moon and returning him safely to the earth's" in an address to Congress on 25 May 1961. It was the third US human space flight programme to fly, preceded by the two-person project Gemini conceived in 1961 to extend space flight capability in support of Apollo.

Kennedy's goal was accomplished on the Apollo-11 mission when astronauts Neil Armstrong and Buzz Aldrin landed their Apollo Lunar Module (LM) on 20 July 1969, and walked on the lunar surface, while Michael Collins remained in lunar orbit in the Command and Service Module (CSM), and all three landed safely earth on 24 July. Five subsequent Apollo missions also landed astronauts on the moon, the last, Apollo-17, in December 1972. From these six spacecrafts twelve people walked on the moon.

Apollo program ran from 1961 to 1972, with the first crewed flight in 1968. It encountered a major setback in 1967 when an Apollo-11 cabin fire killed the entire crew during a prelaunch test. After the first successful landing, sufficient flight hardware remained for nine follow-on landings with a plan for extended lunar geological and astrophysical exploration. Budget cuts forced the cancellation of three of these. Five of the remaining six missions achieved successful landings, but the Apollo-13 landing was prevented by an oxygen tank explosion in transit to the moon, which destroyed the service module's capability to provide electrical power, crippling the CSM's propulsion and life-support systems. The crew returned to the earth safely by using lunar module as a "lifeboat" for these functions. Apollo used the Saturn family of rockets as launch vehicles, which were also used for an Apollo applications programme, which consisted of Skylab, a space station that supported three crewed missions in 1973-74, and the Apollo–Soyuz Test project, a joint United States–Soviet Union low earth orbit mission in 1975. Apollo set several major human spaceflight milestones. It stands alone in sending crewed missions beyond low earth orbit. Apollo-8 was the first crewed spacecraft to orbit another celestial body and Apollo-11 was the first crewed spacecraft to land humans on the moon. Overall, the Apollo programme returned 842 pounds (382 kg) of lunar rocks and soil to earth, greatly contributing to the understanding of the moon's composition and geological history. The programme laid the foundation for NASA's subsequent human spaceflight capability and funded the construction of its Johnson Space Center and Kennedy Space Center. Apollo also spurred advances in many areas of technology incidental to rocketry and human spaceflight, including avionics, telecommunications, and computers.

Japan's Moon Missions

The (Japanese) lunar exploration programme is a programme of robotic and human missions to the moon undertaken by the Japanese Aerospace Exploration Agency (JAXA) and its division, the Institute of Space and Astronautical Science (ISAS). It is also one of the three major enterprises of the JAXA Space Exploration Center (JSPEC). The main goal of the programme is "to elucidate the origin and evolution of the moon and utilize the moon in the future.

The first spacecraft of the programme, the unmanned lunar orbiter SELENE (Kaguya), was launched from the Tanegashima Space Center of 14 September 2007, after being delayed several times.

SELENE-2, Japan's first lunar lander and rover, was expected to be launched in the 2020s, but the mission was cancelled in March 2015. The programme also included a lunar sample return mission (SELENE-3) a mission to Mars to collect data for future manned expeditions (MELOS), participation in the Mars International Sample Return Mission and an advanced lander for future human missions to the moon. The eventual goal is to participate in an international lunar outpost programme, in which Japanese crews would stay on the lunar surface for a prolonged period of time and promote scientific research and environment utilization.

SELENE (KAGUYA)

Setenological and Engineering Explorer (SELENE), nicknamed Kaguya after a lunar princess in the ancient Japanese folklore "The Tale of the Bamboo Cutter", is the second Japanese mission to the moon, launched in September 2007, it was "the largest lunar mission since the Apollo programme". The mission featured three separate spacecraft, the main orbiter (Kaguya), the small relay satellite (Okina), and the VLBI satellite (Ouna).

SELENE-2

SELENE-2, also known as Kaguya-2, formerly known as SELENE-B, would have been the follow-on mission to Kaguya. SELENE-2 would have consisted of one large lander, which would have a small-sized rover, and, if possible, would also have some penetrators inherited from the Lunar-A mission, and a small data relay satellite. In March 2015, JAXA cancelled the SELENE-2 mission.

SELENE–3

SELENE-3 aimed to return about 100 gm of samples from the moon's surface. It would have been launched around 2020. However, due to financial issues, this mission was considered for merging into the option-3 of SELENE-X.

SELENE-X

SELENE-X would have been launched in the late 2010s, in view of Japan's participation in humans lunar activities foreseen.

The SELENE-X may perform either of the following demonstrations:

Option-1: Technology demonstration for building outposts such as the excavation for construction of infrastructures.

Option-2: Logistics capability demonstration for building common landers for both transportation and JAXA's own robotic missions.

Option-3: Highly sophisticated *in situ* robotic lander or returning samples of the surface soil to the earth, including the development of high-speed re-entry capsules.

There are other options under study and will be determind after the international exploration strategy has been clarified.

China's Moon Missions

The Chinese Lunar Exploration Programme (CLEP), also known as the Change Project after the Chinese moon-Goddess Chang'e, is an ongoing series of robotic moon missions by the China National Space Administration (CNSA). The programme incorporates lunar orbiters, landers, rovers, and sample return spacecraft, launched using long march rockets. Launches and flights are monitored by a Telemetry, Tracking, and Command (TT&C) system, which uses 50-m (160 ft) radio antennas in Beijing and 40-m (130 ft) antennas in Kunming, Shanghai, and Urumqi to form a 3,000-km (1,900 mile) VLBI antenna. A proprietary ground application system is responsible for downlink data reception.

Ouyang Ziyuan, a geologist, chemical cosmologist and the programme's chief scientist, was among the first to advocate the exploitation not only of known lunar reserves of metals such as titanium, but also of helium-3, an ideal fuel for future nuclear fusion power plants. Scientist Sun Jiadong is the programme's general designer and Sun Zezhou is deputy general designer. The leading programme manager is Luan Enjie.

The first spacecraft of the programme, the Chang'e-1 lunar orbiter, was launched from Xichang Satellite Launch Center on 24 October 2007, having been delayed from the initial planned date of 17-19 April 2007. A second orbiter, Chang'e-2, was launched on 1 October 2010. Chang'e-3, which includes a lander and a rover, was launched on 1 December 2013 and successfully soft-landed on the moon on 14 December 2013, Chang'e-4, which includes a lander and a rover, was launched on 7 December 2018 and landed on 3 January 2019 on the South Pole-Aitken Basin. On the far side of the moon, a sample return mission Chang'e-5, which was launched on 23 November 2020 and returned on 16 December in the same year brought 1,731 g (61.1 oz) of lunar samples to earth.

As indicated by the official insignia, the shape of a calligraphic nascent lunar crescent with two human footprints at its centre, reminiscent of the Chinese character, the Chinese character for "moon", the ultimate objective of the programme is to pave the way for a crewed mission to the moon. CNSA head Zhang Kejian announced that China is planning to build a scientific research station on the moon's South Pole "within the next ten years (2019-29)".

India's Moon Missions

The Indian Space Research Organisation (ISRO) has carried out 111 spacecraft missions, seventy-nine launch missions, and planned several missions including the Aditya, Gaganyaan, and MOM2.

LUNAR PROGRAMME

Chandrayaan-1 was the first Indian lunar probe under the Chandrayaan programme. It was launched by the ISRO in October 2008 and operated until August 2009. The mission included a lunar orbiter and an impactor. India launched the spacecraft using a PSLV-XL rocket on 22 October 2008 at 00:52 UTC from the Satish Dhawan Space Centre, Sriharikota, Andhra Pradesh. The mission was a major boost to India's space programme, as India researched and developed its own technology in order to explore the moon. The vehicle was inserted into lunar orbit on 8 November 2008.

On 14 November 2008, the Moon Impact Probe separated from the Chandrayaan orbiter at 14:36 UTC and struck the South Pole in a controlled manner, making India the fourth country to place its flag insignia on the moon. The probe hit near the crater Shackleton at 15:01 UTC, ejecting subsurface soil that could be analysed for the presence of lunar water ice. The location of impact was named Jawahar Point.

The estimated cost for the project was Rs. 386 crore. It was intended to survey the lunar surface over a two-year period to produce a complete map of the chemical composition at the surface and three-dimensional topography. The polar regions are of special interest as they might contain water ice. Among its many achievements was the discovery of widespread presence of water molecules in lunar soil.

After almost a year, the orbiter started suffering from several technical issues including failure of the star tracker and poor thermal shielding; Chandrayaan-1 stopped communicating at about 20:00 UTC on 28 August 2009, shortly after which the ISRO officially declared that the mission was over. Chandrayaan-1 operated for 312 days as opposed to the intended two years, but the mission achieved most of its scientific objectives.

On 2 July 2016, NASA used ground-based radar systems to relocate Chandrayaan-1 in its lunar orbit, more than seven years after its shut down. Repeated observations over the next three months allowed a precise determination of its orbit which varies between 150 and 270 km (93 and 168 mile) in altitude every two years.

CHANDRAYAAN-2

Chandrayaan-2 is the second lunar exploration mission developed by the ISRO, after Chandrayaan-1. It consists of a lunar orbiter, also included the Vikram lander, and the Pragyan lunar rover, all of which were developed in India. The main scientific objective is to map and study the variations in lunar surface composition, as well as the location and abundance of lunar water.

The spacecraft was launched on its mission to the moon from the second launch pad at the Satish Dhawan Space Centre on 22 July 2019 at 09:13:12 UTC by a GSLV, Mark III-M1. The craft reached the moon's orbit on 20 August 2019 and began orbital positioning manoeuvres for the landing of the Vikram lander. The lander and the rover were scheduled to land on the near side of the moon, in the south polar region at a latitude of about 70° south on 6 September 2019 and conduct scientific experiments for one lunar day, which approximates to two earth weeks. A successful soft landing would have made India the fourth country after the Luna-9 (Soviet Union), Surveyor-1 (United States), and Chang'e-3 (China) to do so.

However, the lander crashed when it deviated from its intended trajectory while attempting to land on 6 September 2019. According to a failure analysis report submitted to ISRO, the crash was caused by a software glitch. ISRO planned a re-attempt a landing in 2022 with Chandrayaan-3.

CHANDRAYAAN-3

In November 2019, ISRO officials stated that a new lunar lander mission is being studied for launch in April–June 2021; this new proposal is called Chandrayaan-3 and it would be a re-attempt to demonstrate the landing capabilities needed for the Lunar Polar Exploration Mission proposed in partnership with Japan in 2024. If funded, this re-attempt would not include launching an orbiter. The proposed configuration would have a rover. According to VSSC director, S. Somanath, there will be more follow-up missions in the Chandrayaan programme.

According to *the Times of India*, work on Chandrayaan-3 commenced on 14 November 2019. In December 2019, it was reported that ISRO requested the initial funding of the project, amounting to Rs. 75 crore, of which Rs. 60 crore is intended for machinery, equipment and other capital expenditure, while the remaining Rs. 15 crore is sought under revenue expenditure head. Confirming the existence of the project, K. Sivan stated that its cost would be around Rs. 615 crore.

Chandrayaan-3 was a planned third lunar exploration mission by ISRO. Following Chandrayaan-2, where a last-minute glitch in the soft-landing guidance softwere led to the failure of the lander's soft-landing attempt after a successful orbital insertion, another lunar mission demonstrating soft landing was proposed. Chandrayaan-3 was a mission repeat of Chandrayaan-2, but only included a lander and a rover similar to that of Chandrayaan-2, it did not have an orbiter. The spacecraft was planned to launched in the third quarter of 2022. However it was launched from the Satish Dhawan Space Centre on 14 July 2033. The mission was successful entering the lunar orbit on 5 August and India become the first country to touch down lunar South Pole on 23 August 2023, making it the fourth country to land on the moon.

Former prime Minister Atal Bihari Vajpayee announced the Chandrayaan project in his Independence Day speech on 15 August 2003. The mission was a major boost to India's space programme. The idea of an Indian scientific mission to the moon was first mooted in 1999 during a meeting of the Indian Academy of Sciences. The Astronautical Society of India carried forward the idea in 2000. Soon after, the ISRO set up the National Lunar Mission Task Force which concluded that ISRO has the technical expertise to carry out an Indian mission to the moon. In April 2003, over 100 eminent Indian scientists in the fields of planetary and space sciences, earth sciences, physics, chemistry, astronomy, astrophysics, engineering, and communication sciences discussed and approved the Task Force recommendation to launch an Indian probe to the moon. Six months later, in November, the Indian government gave the nod for the mission.

Moon Mission of European Space Agency

SMART-1 is a technology demonstration mission for deep-space solar electrical propulsion and technologies for the future. SMART-1 is the Europe's first lunar mission and it will contribute to developing an international programme of lunar exploration.

The spacecraft was launched on 27 September 2003, as an auxiliary passenger to GTO on Ariane-5, to reach the moon after fifteen month's cruises, with a lunar capture due on 17 November 2004 (one year exactly after the very memorable International Lunar Conference in Hawaii). SMART-1 carried seven experiments, including three remote-sensing instruments that were used during the mission's nominal six months in lunar orbit. These instruments were to contribute to key planetary scientific questions, related to theories of lunar origin and evolution, the global and local crustal composition, the search for cold traps at the lunar poles and the mapping of potential lunar resources. The spacecraft entered the lunar orbit on 15 November 2004 and on fulflling its mission, on 3 September 2006 it was crashed into the surface of the moon.

20

Satellites: An Overview

A SATELLITE is an object in space that orbits or circles around a bigger object. There are two kinds of satellites: natural (such as the moon orbiting the earth) or artificial (such as the international space station orbiting the earth).

Types of Satellites and Functions

1. Communications satellite,
2. Remote-sensing satellite,
3. Navigation satellite,
4. Geocentric orbit type satellites (LEO, MEO, and HEO),
5. Global Positioning Systems (GPS),
6. Geostationary Satellites (GEOS),
7. Drone satellites,

8. Ground satellites,
9. Polar satellites, and
10. Nano satellites, cubesats, and smallsats.

Satellites can be classified by their functions since these are launched into space for specific purposes. The satellite must be designed specifically to fulfil its role. There are ten different types of satellites as listed above.

Communication satellites are artificial satellites that receive signals from an earth station and then retransmit the signals to other earth stations. They commonly move in a geostationary orbit. A remote sensing instrument collects information about an object.

Satellites provide in-flight phone communications on airplanes and are often the main conductor of voice communication for rural areas and areas where phone lines are damaged after a disaster. Satellites also provide the primary timing source for cell phones and pagers.

There are nearly 6,542 satellites orbiting the earth as on 1 January 2021. Out of which 3,372 satellites are active and the rest are inactive.

According to UNOOSA records, there are 8,261 satellites orbiting the earth as on January 2022, out of which only 4,852 satellites are active and the rest are inactive.

There are 11,330 individual satellities orbiting the earth at the end of June 2023, a whopping 37.94 per cent rise since January 2022.

As of November 2024, there are 6,764 satellites in the sky, though only 6,714 are currently working according to astronomer Jonathan (11 December 2024).

There have been no observed collisions between natural satellites of any solar system planet or moon. From the observations of the objects, making up the rings of Saturn, it is believed that satellites continually collide and aggregate with each other, leading to debris with limited size constrained to a thin plane.

In January 2020, two different satellites came close and passed within 50-100 ft distance of each other without colliding. Two satellites might collide at 32,000 mph over the Arctic today.

Most satellites do not speak with each other directly, instead, they utilize radio frequency communications with a ground station to relay communications between satellites. The Pentagon wants help for its satellites to talk to each other.

Satellites can travel in pairs. The spacecraft that visit the ISS often orbit near the stations for several hours before docking. Further, there are several satellites that travel in pairs or trios with other satellites. This is because the spacecraft must be in the same orbital plane to travel together.

Common types of satellites include military and civilian earth observation satellites. There are communication satellites, navigation satellites, weather satellites, space telescopes, space stations, and human spacecraft in the orbit.

Number of satellites launched by the differernt countries of the world is as under:

Algeria	06	Chile	02
Arab SCO	15	China, Peoples Republic of	806
Argentina	37	China/Brazil	03
Asiasat	08	Czech Republic	03
Australia	32	Denmark	09
Azerbaizan	02	Des Satellites Societe Europeenne	69
Bangladesh	01		
Belarus	01	Ecuador	02
Bolivia	01	Egypt	11
Brazil	20	Estonia	01
Bulgaria	06	European Space Agency	96
Chechia	02	European TCSO	57

Europen Organisation	10	Netherlands	08
France	109	New ICO	01
France/Germany	02	New Zealand	02
France/Italy	02	Nigeria	05
Germany	76	North Korea	01
Globalstar	84	O 38 Networks	20
Greece	03	Orbcomm	41
India	108	Osingle	38
Indonesia	19	Pakistan	08
INMARSAT	20	Peru	02
Iran	04	Philippines	02
Iraq	01	Poland	04
Israel	20	Portugal	03
ISS	05	Qatar	01
Italy	50	Republic of Rwanda	06
ITSO	87	Republic of Slovenia	02
Japan	208	Republic of Tunisia	01
Kazakhstan	08	RFSCO	02
Kenya	01	Saudi Arabia	15
Kuwait	01	Sea Launch	01
Laos	01	Singapore	18
Lithuania	06	Singapore/Taiwan	02
Luxembourg	13	South Africa	06
Malaysia	09	South Korea	41
Mexico	12	Spain	42
Morrocco	04	Sweden	13
NATO	08	Taiwan	16

Thailand	12	USA	8,832
Turkey	25	USA/Brazil	01
United Arab Emirates	13	USSR	1,544
United Kingdom	658	Venezuela	03
Uruguay	01	Vietnam	03

Uses of Satellites

There are so many satellites in space, because of the fact that it is all about high speed Internet access instead of being constrained

by wires and cables. Satellites can beam Internet access down to the ground from space. And if we have lots of them in orbit, it means the most remote regions can also get connectivity.

The Indian Remote Sensing Satellites are state-of-the-art satellites, indigenously designed and put into polar orbit for useful remote-sensing applications. It was commissioned with the launch of IRS-1A in 1988. IRS is the largest civilian remote-sensing satellite constellation in the world.

Disadvantages of Having too Many Satellites in Space

1. Costs are prohibitive. Satellites are expensive.
2. Signal Reception can be spotty. They send unreliable signal.
3. Propagation delay is a problem.
4. There are no repair shops in space.
5. Satellites are bad for the environment. Sky pollution is caused due to so many satellites. By reflecting the light of the sun, satellites cause streaks of light across the sky where astronomers would prefer darkness for peering into the heavens and where every day people will be robbed of their own views of a dark sky.
6. More satellites orbiting the earth also means there are more chances of collisions and more "space junk" floating around the earth when the satellites stop working. There are no rules

about who can launch satellites, or how many can orbit the earth at one time.

7. If a satellite is thrown out too fast, it will escape from the earth's orbit because the gravitational pull is not sufficient to provide the required centripetal force. With the correct launch speed, the satellite continues in its falling orbit around the earth.

Satellites on the Basis of Orbiting Path

Low Earth Orbit Satellites: These are functional in an elliptical orbit of about 160–600 km.

Sun Synchronous Satellites: These orbits allow a satellite to pass over a section of the earth at the same time of the day.

Geosynchronous Satellites: Also known as Geostationary satellites, which orbit in elliptical orbit of 36,000 km altitude. These revolve with the same speed as the rotation of the earth, hence from the point of observation they appear to be stationary.

Highly Elliptical Orbit Attaining Satellites. These are positioned at the altitude of 504 km to 39,834 km.

Chandrayaan 3 landed on moon

India's Launched Satellites

ARYABHATTA

It was India's first satellite launched by the Soviet Union on 19 April 1975 from Kapustin Yar using a Cosmos-3M launch vehicle. It was built by the ISRO to gain experience on building and operating a satellite in space.

BHASKAR

Bhaskara-I and Bhaskara-II satellites were built by the ISRO and formed India's first low orbit earth observation satellites.

Bhaskar-I was launched on 7 June 1979, from Kapustin Yar on board the Cosmos launch vehicle.

Bhaskar-II was the improved version of Bhaskar-I. It was launched in 1981 from the former USSR. It provided vital inputs for the Indian remote-sensing programme.

ROHINI

Rohini series consisted of four satellites, viz. Rohini Technology Payload (RTP), RS-1, RS-D1, and RS-D2 which were launched by the ISRO's Satellite Launch Vehicle (SLV) in the year 1980 and three of which made it successfully to orbit.

The series was mostly experimental satellites launched on the experimental launch vehicle, SLV.

ARIANE PASSENGER PAYLOAD EXPERIMENT (APPLE)

It was an experimental communication satellite with a C-band transponder launched on 19 June 1981, by Ariane, a launch vehicle of the European Space Agency (ESA) from Kourou in French Guiana.

It was India's first 3-axis stabilized experimental geostationary communication satellite.

STRETCHED ROHINI SATELLITE SERIES (SROSS)

These satellites were developed by ISRO as follow on to the Rohini satellites for conducting astrophysics, remote sensing, and upper atmospheric monitoring experiments as well as for new and novel application-oriented missions.

EXPERIMENTAL SATELLITES

ISRO launched many small satellites mainly for experimental purposes, which include remote sensing, atmospheric studies, payload development, orbit controls, recovery technology, etc. Indian experimental satellites are noted in Table 20.1.

Table 20.1: Indian Experimental Satellites

Satellites	*Launch Date*
Aryabhatta	19 April 1975
RTP	10 August 1979
RS-1	18 July 1980
Apple	19 June 1981
SRE-1	10 January 2007
ANUSAT	20 April 2009
EDUSAT	12 July 2010
YOUTHSAT	20 April 2010
SRMSAT	12 October 2011
Jugnu	12 October 2011

21

Detection of Evidence for the Existence of New Planets

Prolegomenon

SPIRALS of dust around young stars pave the way for the evidence and detection of the new planets. A group of astronomers is proposing that huge spiral patterns observed around some new-born stars, merely a few million years old (about 1 per cent of our sun's age), may be the evidence for the presence of giant unseen planets. Such an idea goes a long way for the observation and detection of new planets. Also this could enable us enrich our knowledge of the early formative phase of a planet.

Though astronomers have listed a number of planets orbiting other stars, the very earliest phases of planet formation are elusive because nascent planets are born and embedded inside vast, pancake-shaped discs of dust and gas encircling new-born stars, known as circumstellar discs.

On the other hand, scientists have found evidences of a planet on the fringes of our solar system tentatively named "Planet Nine". After Pluto was demoted to a dwarf planet, this could be a major breakthrough in our understanding of our cosmic home. Michael E. Brown, one of the astronomers responsible for Pluto's demotion, and Konstantin Batygin, California Institute of Technology, have come up with an evidence for the existence of this new planet. With computer aided simulations, Brown and Batygin have formulated a theory for the presence of a planet in the undiscovered outer realm of icy bodies beyond Pluto known as the Kuiper Belt, throwing open the possibility that our solar system may be even bigger, and even more populated than we know. The new planet is likely to be smaller than Neptune.

Theory of Gas-and-Dust Discs for the Detection of New Massive Planets

The presence of new planets may be detected through the modification of circumstellar discs on large scales. This is based on detailed computer modelling of how gas-and-dust discs evolve around new-born stars, which was conducted by two NASA Hubble Fellows, Roubing Dong of Lawrence Berkeley National Laboratory, and Zhaohuan Zhu of Princeton University. Their research has appeared in the 5 August 2017 edition of the *Astrophysical Journal Letters.*

According to Dong:

> It is not so easy to observe suspected planets inside a bright disc surrounding a young star. Based on his study, we are convinced that planets can gravitationally excite structures in the disc.

Another method of planet detection is the identification of large-scale features produced by planets. This method is quite different from all other techniques presently used. This approach can go a long way in answering many questions of astronomical observers.

Gaps and rings observed in other circumstellar discs suggest invisible planets embedded in the disc. It is also difficult to ascertain the presence of a new planet from the observed gap, because the gaps may be created by a planet's gravity. Also, multiple planets together may open a single common gap. So, it is very difficult to estimate their number and masses.

Two large-scale spiral arms around two young stars, SAO 206462 and MWC 758, have been photographed by the telescopes mounted on the ground of the earth. A few other nearby stars also show smaller spiral-like features. Astronomers are now busy in searching for the causes of the creation of such spirals around some stars.

The first planet orbiting a normal star was identified in 1995. with the help of ground-based telescopes and NASA's Kepler Mission, a number of exoplanets have been catalogued to data. Since the planets are in mature systems, many millions or a few billion years old, they offer little direct clues as to how they formed.

Dong has further added that there exist many theories about the formation of the planets, but very little work is based on the direct observational evidence confirming these theories. Using NASA's James Webb Space Telescope, the astronomers can trace the circumstellar discs and look for features, as simulated by the modelling. This will help them directly observe the predicted planet causing the density waves. The natural gravitational interaction between the disc and the planet creates regions where the density of gas and dust increases.

According to the scientist Zhu Jin simulations also suggest that these spiral arms can provide rich information about an unseen planet, revealing not only its position but also its mass.

22

Aditya-1

India's Maiden Mission to the Sun

Introduction

In our solar family, the sun is the primordial star, which controls the entire solar system. The earth and other planets in the solar system get light from the sun and afterwards radiate the same. A habitable world and barren wasteland are distinguished from each other due to the difference in getting the solar energy. Very recently we were trying to understand the various dynamics of our parent star. Solar energy shapes the life on earth. The physical processes undergoing on the surface of the earth depend on solar energy and our food chain of ours is sustained by the sun. Earth's climate is governed by the sun and its variability is a hotly debated issue in climate science. For instance, the little Ice Age of 1816 was the result of a mere 1° fall in global temperature. According to Bill Bryson, 1816 came to be known as the year without summer. Crops failed to grow. In Ireland, a famine and a subsequent typhoid epidemic killed around 65,000 people. Ostensibly, the countries which receive less solar energy become cold and those getting more sunlight become hot. Accordingly, the weather of the countries varies. Climatic changes occur on the earth due to variations of solar energy received by our planet.

Owing to thermonuclear reactions, energy is being continuously generated inside the sun. The sun consists chiefly of hydrogen, the lightest and simplest of all elements. Mass is converted into energy in the sun due to Einstein's famous relation $E = mc^2$, where c is the speed of light in vacuum.

The above knowledge was put to fruitful use by Hans Bethe, Professor of Physics at Cornell University, USA, who advanced the theory in 1938 that the extremely high temperatures found in the sun and the enormous pressure exerted by its great bulk were propitious to cause fusion of hydrogen nuclei producing helium nuclei, resulting in a loss of mass and release of energy. This is also the reaction in the hydrogen bomb.

Man has been curious about the sun, looking at it with wonder and awe ever since he became aware of the surroundings. In order to study the sun more and for knowing the activities going inside the sun, man has started solar mission now and then. This is certainly a new awakening germinated in man. Different nations of the world are continuously trying through their solar voyages to understand sun's physical and chemical changes time and again. More than twenty international space missions are launched to understand the sun's structure and composition. Let us look at India's effort in this domain.

Aditya-1 Mission

Today Aditya-L1 spacecraft has completed its first halo orbit around the sun–earth L-1 point. The Aditya-L1 mission is an Indian solar observatory at Lagrangian point L1, launched on 2 September 2023 and was inserted in its targetted halo orbit on 6 January 2024 (view on 17 November 2024) Earlier USA, Japan, and European Space Agency have sent missions to the sun. After launching Aditya-1, India has joined the select group.

Components of Aditya-1

ASPEX: This part of Aditya-1 will study the variation, distribution, and spectral characteristic of solar wind, which can affect our power lines, communication satellites, and high altitude spacecraft.

VELC: This component will study the parameters of the solar corona and origin of Coronal Mass Ejections (CMEs), which can collide with earth's magnetic field and change its shape.

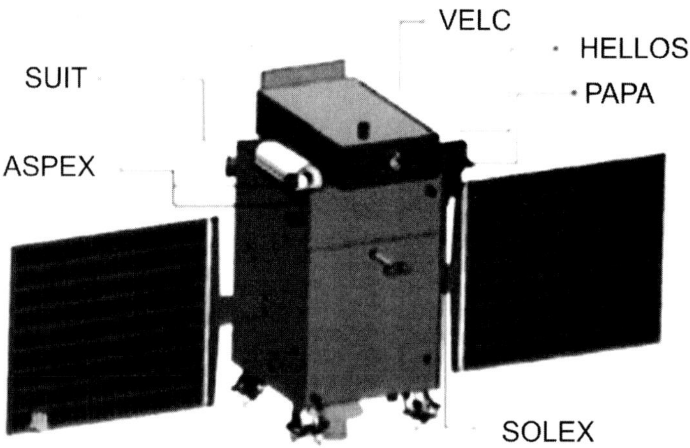

*fig.*22.1: Prototype of India's Aditya-1: its various components

SUIT: This part will image the photosphere and chromosphere in UV range. A better understanding can help us keep track solar flares emanating from the photosphere.

SOLEX: This part will monitor X-ray flares to study the heating mechanism of the corona. Energy from X-ray flares can disrupt radio waves, causing blackouts in navigation and communications signals.

PAPA: This component of Aditya-1 will study the composition of solar wind and its energy distribution. Solar wind can disrupt communication and navigation satellites.

HELIOS: This part will observe the dynamic events in the corona and estimate the energy used to accelerate the particles during the eruptive events. An estimate of energy can help us shield ourselves in an effective and timely manner.

Aims and Objectives of Aditya-1 Mission

Indian Space Research Organisation (ISRO) is leading the maiden Aditya-1 Mission. According to Deviprasad Karnik a scientist at ISRO, the primary objective of this mission is to study the solar corona, processes leading to changes in it, and to understand what causes heat in the corona. Observations of the sun's photosphere, chromospheres and corona are also possible.

The corona is the outermost layer of the sun's atmosphere, preceded by the chromosphere and photosphere respectively. The mission will provide a multi-pronged holistic approach to understand some of the outstanding problems of solar physics.

The design of payloads is being processed. Aditya-1 will be launched from Sriharikota, an island off Sullurupeta, a small town in Nellore district, Andhra Pradesh.

Besides ISRO, many other institutes are closely associated with this mission for studying the sun. The Indian Institute of Astrophysics, Bengaluru, is working on the visible Emission Line Coronagraph. The Physical Research Laboratory, Ahmedabad, is working on the Aditya Solar Wind Particle Experiment. The Vikram Sarabhai Space Centre, Thiruvananthapuram, is working on Plasma Analyser Package for Aditya.

Initial Concept and Further Developments

In 2008, launching of Aditya-1 Mission was conceptualized. Its launch was envisioned to coincide with solar maxima – a period of intense sun activity that occurs every eleven years. However, it was later decided that the satellite would be placed in a halo orbit around lagrangian point 1, which is 1.5 million km from the earth (where a satellite can maintain its position with respect to other bodies). According to Karnik, the initial concept was to put the coronagraph payload in a small satellite in an 800 km low earth orbit. Aditya-1 is an updated version of the original Aditya Mission with six additional payloads.

Placing the satellite in a halo orbit has many advantages. Situated outside the earth's atmosphere and magnetosphere, it is not affected by the earth. A halo orbit also ensures that there is no occultation for the spacecraft's line of sight. A satellite in this position experiences much less mechanical and thermal disturbances and hence it becomes more stable.

Results of Past Missions

Earlier sun missions provide us many information about the sun. The Solar and Heliospheric Observatory (SOHO), a collaboration between the European Space Agency and the National Aeronautics and Space Administration (NASA), celebrated its twentieth anniversary on 2 December 2015. Bernhard Fleck, a project scientist with SOHO, is of the opinion that SOHO has changed our notions about the sun in many ways. It has opened up our eyes in understanding the sun better.

In course of time, many solar observatories came up to study the sun. These are:

1. Hinode, the Solar Dynamics Observatory (SDO),
2. The Interface Region Imaging Spectrograph, and
3. The Solar and Terrestrial Relations Observatory.

NASA's SDO has been recording the sun's dynamic solar activity since its launch on 2 November 2011. The objective of SDO mission was to understand the life cycle of the solar magnetic field.

It is highly necessary to predict solar activities for the safety of lives and properties on the earth. At times, solar flares occur. On the night of 2 September 1859, the world witnessed its first recorded solar flare, which is also known as solar storm. Recently a solar storm was predicted on 23 July 2012. Fortunately, we missed it by a week, because of the fact that the earth had moved ahead in its orbit by the time.

Frontiers of Investigations on Solar Activities

Space weather can be affected by the disruptions in the solar atmosphere and surface. Aeronautics, astronautics, and technology infrastructure can be advanced by understanding the sun more clearly. According to Dean Pesnell, the next frontier in our understanding of the sun is to measure its polar regions with the same instrumentation employed near the equator. Presently, a few measurements above the poles have been made.

Pesnell further views that the better predictions will lead to a more cost-effective response. Power plants that will be affected can be isolated. Satellites can be tuned to power off sensitive high-voltage components, and astronauts in deep space can get into a safe shelter before the danger arrives.

In addition to the above prospective, Aditya-1 can also reveal the cause of heating up the solar corona to a high temperature, a million degrees or so, much more than the visible outer layer of the sun's surface, the photosphere.

Aditya-1 is different from previous missions to the sun. It is a multi-wavelength observatory that will look at different layers of the solar atmosphere. Aditya-1 is different from NASA's STEREO which has two coronagraphs and one imager. The coronagraphs on board STEREO take images every 10 minutes and probes only the outer corona. Aditya-1's VELC, on the other hand, will look at the inner corona and will take images every second. Thus, Aditya-1 mission is more promising than the earlier solar missions.

23

Scramjet Technology in Indian Rocketery

Prolegomenon

A rocket is based on the recoil principle. Fuel and oxygen are burnt in the ignition chamber. As hot gases escape from a rear opening with some backward momentum, the rocket moves in the forward direction.

Obviously the rocket works on the principle of conservation of momentum. Momentum of the rocket before explosion of gas is zero, and, after explosion, the sum of forward and backward momentum also becomes zero. As a result the momentum of the rocket is conserved.

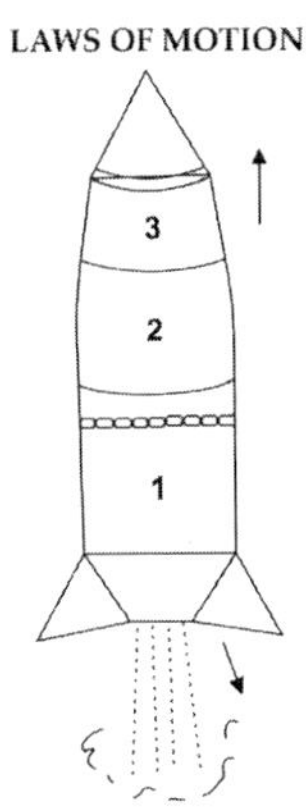

*fig.*23.1: Motion of a rocket

From the above explanation, it is well understood that the propulsion of a body takes place by means of a force produced by discharging a fluid in the form of a jet. The backward moving jet fluid reacts on the body in which it was produced; in accordance with Newton's Third Law of Motion, to create a reactive force that drives the body forward.

Jet propulsion occurs in nature, the squid using a form of it to propel itself through water. Although jet-propelled boats and cars have been developed, the main use of jet propulsion is in aircraft and spacecraft. Jet propulsion is the only known method of propulsion in space. In the atmosphere, jet propulsion becomes more efficient at higher altitudes, as efficiency is inversely proportional to the density of the medium through which a body is flying. The three principal means of providing jet propulsion are the turbojet, the ramjet, and the rocket.

The turbojet is an air-breathing heat engine based on the gas turbine, used to power the jet aircraft.

The ramjet is also an air-breathing engine, but compression of the oxidant is achieved by the forward motion of the device through the atmosphere. This enables the compressor and turbine of the gas turbine to be dispensed with and the remaining system consists simply of an inlet diffuser, a combustion chamber in which fuel is burnt, and a jet nozzle through which the products of combustion are discharged, used in guided missiles, the ramjet must be accelerated to its operating velocity before it can fly.

These two forms of jet propulsion, being air-breathing engines, can only be used in the earth's atmosphere. The rocket, however, carries its own oxidant and can thus be used in space.

Scramjet

A scramjet, which actually is a "supersonic combusting ramjet", is a variant of the ramjet and a very advanced version of the air-breathing jet engine. Both ramjet and scramjet engines are called air-breathing rocket engines, as they suck and compress the air from the atmosphere for combustion. Ramjet is an advanced design of ordinary jet engines. In an ordinary jet engine, air is compressed by the rotary blades and brought to the combustion chamber. As the fuel is injected into the hot compressed air, it burns making the whole mixture expand considerably. The exhaust gas comes out of the nozzle generating thrust, as per Newton's Third Law as mentioned earlier.

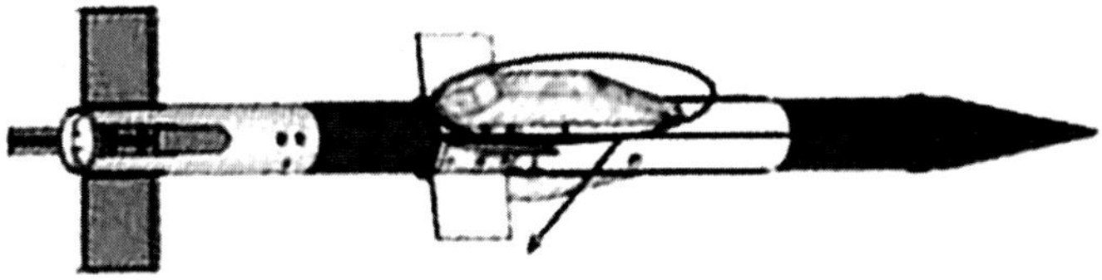

fig.23.2: Scramjet Engine

A ramjet, instead of using rotary blades to ingest compressed air into the combustion chamber, the compressed air is sucked through inlets as the vehicle moves at supersonic speeds. In a typical A ramjet, the converging air inlet in its nose is used to compress the air and in the combustor gaseous fuel is burned with atmospheric

oxygen to produce heat and a diverging nozzle where the heated expanding air is accelerated to generate thrust.

On the other hand, scramjet is an improved version of the ramjet which operates optimally at hypersonic speeds, while ramjet works below Mach 5, i.e. at supersonic speeds.

Mach number is defined as the ratio of the relative speeds of a fluid and a rigid body to the speed of sound in that fluid under the same conditions of temperature and pressure. If the Mach number exceeds 1, the fluid or body is moving at a supersonic speed. If the Mach number exceeds 5, it is said to be hypersonic. The number is named after Ernst Mach (1838–1916).

The description of the scramjet engine is deceptively simple, but its design is really complex. Imagine sitting in the window seat of a fast moving bus. If you put your head out, as the strong wind blows against your face, you feel difficulty in breathing. Likewise, as the scramjet engine lifts off and obtains considerable speed, the compressed air inlet is a technical challenge. Imagine holding a burning candle and running in the open field. The relative wind velocity would surely extinguish the fire. In supersonic conditions inside the scramjet engines sustaining the ignition is a huge challenge. In fact, technologists have compared initiating the ignition with trying to light a matchstick in a storm.

An eminent scientist T.V. Venkateswaran has analysed clearly the functions of ramjets and scramjets. According to such explanations, ramjets or scramjets cannot produce thrust at zero airspeed; they cannot move an aircraft from the standstill position. Only when the rocket achieves supersonic or hypersonic speeds, then the ramjet and scramjet can be ignited. The initial thrust has to be provided with conventional solid fuel rockets. Once a certain speed is obtained, within milliseconds the scramjet engine has to be ignited.

The ramjet work best at supersonic speeds, and its efficiency goes down when the speed reaches hypersonic range, while scramjets work well at hypersonic speeds and are not as efficient at supersonic speeds. Making the best of both the worlds, Indian Space Research Organisation (ISRO) is trying to develop a Dual Mode Ramjet (DMRJ) engine, which uses hydrogen as fuel and air from the atmosphere as the oxidizer, which will function as ramjet in lower Mach speeds and as scramjet once the rate of the rocket increases to more than Mach 5.

Merits of Scramjet Engines

Scramjet engines would reduce the chance of failure considerably, because of the reduction of the number of stages of the rocket launch and having no movable parts. All the current rocket technologies are for one-time use. The launch cost is not only the cost of the payload and fuel but also the launch vehicle.

Further, the efficiency concerning payload to lift off mass is low, of the order of 2-4 per cent. GSLV weighs about 400 tonnes and the payload is only about 6 tonnes. More than 200 tonnes of the weight at liftoff is the oxidizer, which is used up in the first 50 km of the flight. Scramjets use air from the atmosphere to burn fuel, thus saving the need to carry an oxidizer, consequently increasing the payload efficiency of the craft. By enabling reuse of the launch vehicle and reducing the weight at liftoff, the cost of sending the payload to space can be reduced significantly. One estimate shows that, from the current price of US $5,000 per kg, the cost can be brought down to US $2,000 per kg. This makes reaching space affordable and safe.

ISRO's Quest for Scramjet Technology

ISRO conducted successful in-flight test of scramjet engines. Indigenously built scramjet engines weighing 3,277 kg were launched from the Sriharikota Range (SHAR), Satisth Dhawan Space Centre at Sriharikota near Chennai on 28 August 2016. This engine was carrying a two-stage RH-560 sounding rocket. The entire system was hugging on both sides of the Rohini class.

ISRO's maiden attempt to test the air-breathing prolusion system, which used hydrogen as fuel and oxygen from the atmosphere as the oxidizer, was successful, with the propulsion system remaining in desired flight condition of Mach number (6 + 0.5) and dynamic pressure (80 + 35 kilo pascal). It is only under these circumstances that the ignition of the scramjet engine can be activated and the combustor module can be sustained for the rocket launch. These operation parameters are proposed to be tested in future on ground and flight trials. The preliminary examination of the tests indicates that the burnout of booster rocket stage and activation of scramjet engines for 5 seconds between 55 and 60 seconds of the flight at 20-km height, followed by the burnout of the second stage took place as planned. Obviously, with the successful ignition and sustaining the flame for about 5 seconds in the supersonic flight conditions, ISRO has crossed a milestone in its quest for scramjet technology joining the elite club of USA, Russia, and the European Space Agency having expertise in the technology.

Technical Challenges and Future Prospective

It is observed that the scramjets look simple, but designing and developing it is an extremely technical challenge. Air inlet design, compressing the air to desired levels, computing the complex flows of high temperature and highly compressed gases, fuel injection, and sustaining the flame make it tough for designing suitable scramjet engines.

Since scramjet technologies rest on strategic know-how and no country is willing to share its technical knowledge, ISRO is trying to develop its own and for that indigenous programmes have been chalked out. Only recently Computational Fluid Dynamic (CFD) tools have matured and hence it is only by trial and error flight testing that the technology can be designed. Through computational simulation of hypersonic flow and various stages of preparation of a scramjet engine, ISRO can finally achieve one such technology, which is a challenge for the ISRO scientists.

24

Discovery of Gravitational Waves

Introduction

THE prediction regarding the existence of gravitational waves, waited for over 100 years, has now come true. The discovery of the Gravitational Waves (GWs) has opened new channels for understanding our universe in many ways. The possibility of existence of GWs was observed theoretically by Albert Einstein in his general theory of relativity in 1916. Einstein addressed the inadequacy in the Newtonian perception of absolute space and absolute time as well as the instantaneous interaction by gravitational force with space and time. He argued that space and time are neither absolute nor distinct, but are woven into a single space–time manifold. According to Einstein, gravity is not an instantaneous force, rather, it is the curvature in the space–time fabric. Like the accelerated charge-producing Electromagnetic (EM) waves emit electromagnetic energy, similarly, the accelerated mass can produce GWs as ripples in the space–time curvature in emitting gravitational energy.

It is interesting to mark that all the phenomena predicted by the general theory of relativity were experimentally verified except the GWs. Over 100 years, this was bothering the scientists. This is due to the fact that gravity is normally a weak force and the ripples in space–time are so small that these eluded detection of GWS till now. In other words, the amplitudes of the waves are extremely small, just a millionth of the size of an atom. Nevertheless they provide exciting possibilities in the realm of astrophysical investigations including the understanding of the origin of the universe from the Big Bang.

The general theory of relativity was so subtle and complex that its solutions and the necessary experimentations to verify all its predictions have taken a lot of time, although two other predictions like bending of starlight near the sun and precession of planetary orbit like Mercury had been verified experimentally long since. It had not been possible to observe the third most important prediction for the existence of GWs. It had taken almost seventy-five years to think and design an appropriate experimentation technique and twenty-five years to achieve the precession and accuracy in the observation and measurement of GWs. Let us have a cursory look at the prolonged research over the last 100 years for detecting GWs, which has become successful on 14 September 2015 as per the declaration GW discovery on 11 February 2016.

Continued Investigations of GWs

1915: Albert Einstein's General Theory of Relativity came to light.

1916: Einstein predicted the existence of GWs as a consequence of the General Theory of Realtivity (GTR).

1957: Physicist Richard Feynman and mathematician Hermann Bondi used the sticky-bead thought experiment to demonstrate that if GWs did exist, they would theoretically transfer energy, which could be detected.

1962: The first paper, detailing the principles to use laser interferometers for the detection of long GWs, was brought out by Soviet Physicists Michael Gerstenshtein and Vladislav Pustovoit.

1968: Physicist Joseph Weber falsely claimed the detection of GWs. This false claim for the detection of GWs inspired Rainer Weiss, a

German-born American Physicist who won the Noble Prize in 2017, to set up the Laser Interferometer Gravitational-Wave Observatory (LIGO).

Late 1970s: A binary star system, of which one is a pulsar, was observed by two British scientists Russell Hulse and Joseph Taylor in 1974. The characteristic of their orbit was deduced from the Doppler shifting of radio signals given off by the pulsar. These stars are about 1.4 solar mass and the size of their orbit was about 1/75 of the earth–sun orbit. Because of the greater mass and their smaller separation, it involved a greater acceleration support by the two massive bodies resulting in continuous loss of energy by gravitational radiation.

As the energy is given up, the stars draw closer to each other by inspiralling which was observed in the pulsar's signals. The measurements on this Hulse–Taylor system had been carried out over several years to find that the gravitational radiation, predicted by general relativity, matches these observations within 0.2 per cent. For this, Hulse and Taylor were awarded the Nobel Prize in 1993. However, this was not the direct observation of GW.

1984: Ground-based LIGOs were established. LIGO has three detectors: one at Livingston, Louisiana, another at Hanford, Richland, Washington; and a third (formerly installed as a second detector at Hanford, California). India has set to house the next detector of the LIGO in April 2023. The Indian government has approved its construction, and by the end of the decade, they expect to see the observatory's first detection of GWS (view on 13 February 2024).

Each observatory has two light storage arms that are 4 km in length. These are at 90° angles to each other, with the light passing through 1-m diameter vacuum tubes running the entire 4 km. A passing gravitational wave may slightly stretch one arm while shortening the other. This is precisely the motion to which an interferometer is most sensitive.

LIGO–India Project

LIGO–India is planned to set up at Aundha, Nagnath Hingoli district in Maharashtra. In April 2023, the central government has approved the project to build the LIGO in Maharashtra at an estimated cost of Rs 2,600 crore. This project is under investigation and will come to work soon after construction.

2014: The astronomers at the Harvard Smithsonian Center for Astrophysics falsely claimed the detection of GWs

2015 (14 September): Signals for GWs were recorded by both LIGO detectors at Livingston and Hanford with a time difference of 7 milliseconds due to the angle between the two detectors and the source. Analysis of the signal showed that it came from the Southern Celestial Hemisphere, in the rough direction much farther away than the Magellanic Clouds at 1.3 billion light years. The sources were two black holes with masses 29 and 36 solar masses orbiting each other at a fast speed. At the beginning of the signal, the two objects were circling about each other 30 times a second. By the end of 20 milliseconds snatch of the data, the two had accelerated to 250 times a second before the final collision and a violent merger. In the process it released more power than fifty times of all the stars in the observable universe combined. The gravitational wave signal swept upwards in frequency from 35 to 250 Hz. The mass of the new black hole obtained from the merging of the two was sixty-two solar masses. Energy equivalent to three solar masses was emitted as GWs.

2016 (11 February): The LIGO scientific collaboration teams announced that they had directly detected GWs from a pair of merging black holes using the advanced LIGO detectors.

Experimental Arrangement for Detecting GWs in LIGO

Various GW observatories (detectors) in the world are either under construction or in operation such as the advanced LIGO at Hanford and Livingston. The experimental arrangement for the detection of GW in LIGO is described briefly as:

a. A laser beam emitter (laser source).
b. Two plane mirrors.
c. A beam splitter (prism).
d. A photodetector.
e. Light storge arms (tubes).

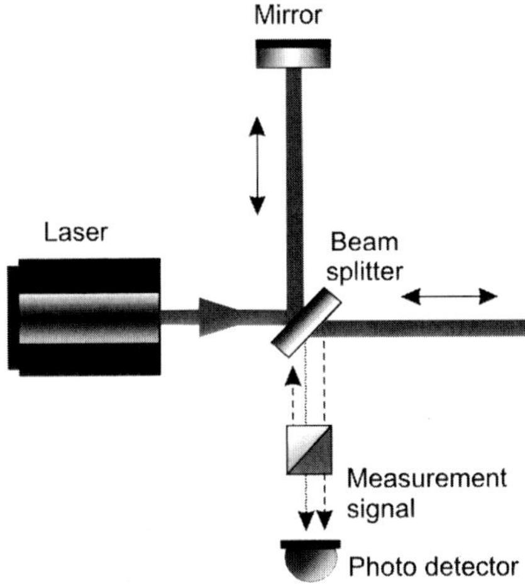

fig. 24.1: LIGO: Experimental arrangement for the detection of GWs

A laser beam is split by a device, called beam splitter, and is sent down in perpendiculary arranged tubes of precisely the same length, each of 4 km. The beam bounces back off the two mirrors at the end of the tube and recombines at the beam splitter. The arrangement is such that the light beams, on recombining, cancel each other out since they interfere destructively. Therefore, no light is detected by the photodetector as long as the arms are of the same length. When gravitational waves pass, the mirrored beams no longer perfectly line up the beams or destructively interfere. Trace amounts of light are detected and measured signalling the detection of GWs.

Future Prospects of the Discovery of GWs

The present discovery of GWs could open new channels of understanding our universe in many ways. By studying sound and seismic waves, we can understand the properties of the world around us. Light and the entire range of the EM spectrum are like the second eye. In some ways, the detection of gravitational waves confirms the way by which the universe operates. In particular, it confirms a long-standing prediction of the general theory of relativity, adding to earlier experimental confirmations.

According to Peter S. Shawhan, Professor of Physics, University of Maryland, USA, we can now track the birth, evolution, and death of stars. He further adds that this discovery is one more confirmation for our understanding of mass and gravity through general relativity, but this measurement does not say anything about other interactions of matter. In future, we will be able to detect things that are not just the mergers of binary black holes and binary neutron stars, and the dark matters and dark energy which constitute almost more than 95 per cent of everything in the universe.

T.V. Venkateswaran, a scientist at Vigyan Prasar, opines that the study of GWs is like opening the third-eye, with which we can understand objects such as black holes and dark matter that do not interact with the EM spectrum. He says the image of the universe has changed into a much more dynamic one, where stars explode and galaxies collide. The expansion of the universe can be also explained by the study of GWs.

Varun Bhalerao, a scientist at the Inter-University Centre for Astronomy and Astrophysics (IUCAA), Pune, says "We are enriching our knowledge of the Cosmos". He is working with the LIGO project.

Varun Bhalerao is an astrophysicist at IIT, Bombay. He obtained a BTech in Electrical Engineering from IIT, Bombay, followed by a PhD from Caltech in 2012. He is the astrophysicist at the Space Technology and Astrophysics Research (STAR) Lab at IIT, Bombay and has worked as an associate professor at IIT, Bombay. He is now working to build cutting edge instruments. He adds:

> We now have a different channel to understand the universe. Gravitational waves are not EM waves. So, we cannot see them through our eyes or with telescopes. But this is the first time ever we have seen black holes of this mass (about 30 times heavier than the sun), and we have seen them collide.

According to Sanjeev V. Dhurandhar, Professor Emeritus at IUCAA,

> Our current knowledge of the universe is almost entirely obtained from EM waves. EM waves are easily scattered or absorbed. On the other hand, GWs are not easily scattered, so it would give high fidelity information. This difference in their properties implies that we would be able to obtain knowledge which was hitherto not available to us.

Conclusion

The detection of GWs is a long-awaited confirmation of Einstein's predictions. This development should lead to the opening up of an entirely new discipline of gravitational astronomy, that will depend purely on GWs, just as the radio waves engendered radio astronomy. Scientists are hopeful at the prospect of such a development as it opens up a new window for understanding the universe, which is yet to be explored.

As suggested by some scientists, we can avail maximum scientific benefits of GWs through a worldwide network with detectors being placed widely across the earth, all scanning the sky together and combining their results.

25

Einstein's Universe

Isaac Newton's theory of gravitation was challenged due to an anomaly in the orbit of Mercury, the planet closest to the sun. Years of observation shows that Mercury's point of nearest approach to the sun, or perihelion, changes from one orbit to the next. It moves forward a bit in each orbit. The anomaly could not be explained by Newton's law. If Newton's law were correct, every planet should follow exactly the same path forever, except for minor irregularities caused by the gravitational influence of other planets in the neighbourhood, as in the case of Uranus. But the change in Mercury's orbit was regular and could not be explained on the basis of the gravitational influence of the other known planets. For a long time, the astronomers searched for the cause of this anamoly, but could not find out a clear-cut answer to it. Finally, in 1915, Albert Einstein, German physicst came out with a solution. In 1905, he published a revolutionary paper on the "special theory of reletivity" which changed our notions about space, time, mass, energy, motion and gravitation. It provided a new approach to study the cosmos. In applying Einstein's theory, the shift of the planet's perihelion could be exactly accounted for. It was one of the first successes of the new theory of gravitation based on relativity, general relativity.

Einstein's theory of general relativity did not agree with Newton's theory in two important aspects. First, Newton considered the mass of a body to be constant, while Einstein's theory asserted the mass of a moving body is by no means constant; it increases with its velocity relative to an observer. Einstein's theory also stated that no material object could ever reach the speed of light, because if tended to do so, its mass would become very large, almost reaching infinity. This theory rejected the idea of gravitation being a force that can be exerted instantaneously over distances, as Newton had assumed in his law of universal gravitation. In Einstein's theory, gravitation was viewed as a property of space than an attractive force between bodies. Further, it did not consider space and time as separate entities, but as a single four-dimensional entity, which Einstein called "space-time".

Einstein's theory did not reject Newton's theory altogether; it only showed that Newton's law does not apply exactly when we deal with speeds approaching the speed of light (over 300,000 km per second), where the theory of relativity is more appropriately applicable. However, for situations that do not involve such fantastic speeds, Newton's law still holds.

Force vs Field

One of the fundamental postulates of Einstein's theory of general relativity is that, over a limited region of space-time, it is impossible for observers to tell whether they are undergoing uniformly accelerated motion or are in a gravitational field. Einstein showed that it was not necessary to think of gravity as the force was at a distance. Instead, he described gravity in terms of its local effects on space and time, i.e. as the curved geometry of space-time, as determined by the distribution of matter and energy.

Simply stated, Einstein's law of gravitation contains nothing about the force. Rather, it describes the behaviour of objects in a gravitational "field". According to Einstein's theory, gravity is a distortion in the fabric of space caused by mass. As a result of the presence of matter, space becomes curved. The stars, planets and other celestial bodies simply follow the lines of least resistance among the curves.

When we talk about the curved space in Einstein's theory, we are actually talking of a four-dimensional entity called "space–time" of which time is an integral part. We cannot visualize Einstein's concept of space–time using any physical analogy because it can be represented only by mathematical equations.

Einstein's general theory of relativity gave a completely new concept of gravitation. It not only provided a plausible solution to the riddle of Mercury's perihelion anomaly, but also predicted two entirely new phenomena that were totally unexpected. First, it predicted that an intense gravitational field should slow down the vibrations of atoms causing a shift of spectral lines towards red. Second, it predicted that a strong gravitational field would bend even light rays. The first effect — the shifting of spectral lines under gravitational field, known as the "Einstein's shift" — was detected in a class of small stars known as white dwarfs in 1975.

The verification of Einstein's second prediction of the bending of light rays by gravitational field came dramatically in 1919. That year, in photographs taken during a total eclipse of the sun, stars were found to indeed shift position when light from them passed very close to the sun. Comparing photographs of the same part of the sky taken earlier, when the sun was not in the vicinity, it was found that stars, which ought to have remained hidden behind the sun during the eclipse, were indeed visible in the photographs. Obviously, light from these stars which were actually behind the sun at the time when the photographs were taken, had been bent by the gravitational field of the sun, which led to their apparent shift in position in the photographs. More significantly, the degree of bending was exactly similar as predicted by Einstein's theory.

Einstein's equations also anticipated the existence of bizarre cosmic objects, such as black holes — remnants of massive stars with gravitational fields so strong that not even light can escape from them. His equations also gave us the famous formula of mass energy equivalence ($E = mc^2$), that is sum the of all actions and creations in the universe. It is an unabated transformation of mass into energy as predicted by Einstein's theory that sustains the stupendously violent processes inside stars and galaxies and makes the cosmos what it is today.

26

Solar Energy

Genesis, Growth and Useful Technical Methods

Introduction

THE total energy of the world is generated from a number of different sources, which is, by convention, broadly classified into two categories, viz. non-renewable and conventional sources of energy. The first category includes fossil fuels, and minerals such as oil, coal, natural gas, and probably uranium as well. The second category embraces solar, wind, geothermal, tidal wave, hydropower and ocean thermal gradient.

The limitation of continued availability from conventional sources has raised a world wide interest in non-conventional sources of enegry. These sources, which are practically inexhaustible, can play a dominant role in energy supplies of the future. All the forms of energy exploited by man, except nuclear energy, are derived directly from the sun. The sun is a star, one of the hundreds of thousands of millions of stars in the universe. Our earth is a planet of our solar system that contains other planets, their satellites, the comets, and meteorites. The sun is the source of living creatures and our very existence depends upon it.

The sun is some 150 million km in diameter. It is a fireball of incandescent (burning) gas with a surface temperature of 5500 °C.

The energy radiated by the sun spreads in all directions and only 5×10^{-10} per cent of total radiation is intercepted by the earth. And between 30-50 per cent of this is only observed by the earth.

Due to its thermonuclear reactions, energy is being continuously generated inside the sun. The central temperature of the sun is about 15×10^6 and the temperature of the photosphere (outer layer) is about 5800 °K. Owing to the high temperature of the sun's core, hydrogen undergoes fusion and the process generates huge amount of energy in the sun.

Energy Generation in the Sun

The sun is considered to be the ultimate source of all forms of energy. It has been supplying huge amount of energy for billions of years and is expected to continue this for many billions of years to come. Then, the question is: how does the sun generate this vast energy?

The answer to this question was first given by Hans Bethe and Carl Weizsacker in 1983. According to them, the interior temperature of our sun is very high (~15 × 10 °C) which is suitable for fusion reactions to occur.

As mostly hydrogen and helium constitute the solar matter, their atoms, under the high temperature of the sun's core, are stripped off their electrons and are in the form of protons (${}_1H^1$) and α-particles (${}_2He^4$). Four hydrogen nuclei fuse to form a single helium nucleus and release 26.7 MeV of energy during the process.

$$4\,{}_1H^1 \longrightarrow {}_2H^4 + 2({}_{+1}e^0) + q.$$

Billions of hydrogen nuclei fuse every second to account for the generation of huge amount of energy in the sun.

In the sun, two types of thermonuclear fusion reactions occur. They are proton–proton cycle and carbon–nitrogen cycle. Both these reactions are more or less responsible for energy generation in the sun.

Technical Methods for Using Sun's Energy

With advances in technology, methods have been designed to utilize the solar power. Solar panels have been devised to conserve solar energy. Now, three methods are available for using solar energy directly or indirectly. These are the mechanical, the photo-electronic, and photosynthetic methods. Air heater, water, and solar cooker work using solar energy under mechanical method. In the photo-electric method, the sun's energy is converted into electric current. By photo-voltaic or photo-electronic method, photo cells work using solar energy. Solar cells were first used in space satellites for providing power to different instruments and communication systems inside the satellite. The first communication satellite, Telstar 1, launched in 1962, was equipped with 5,600 solar cells. The solar cells have now become a standard equipment in all space satellites in preference to other energy devices.

Photosynthic Method: During photosynthesis, plants take in carbon dioxide (CO_2) and water (H_2O) from the air and soil. Within the plant cell, the water is oxidized, meaning it loses electrons, while the CO_2 is reduced, meaning it gains electrons. This tansforms the water into oxygen and CO_2 into glucose.

Photo-Electric Method: This is the process by which light (sun-rays) is converted to electricty, i.e. light energy is converted to electric energy.

Photo-Electronic Method: In this process, if light is incident on a material, electrons are emerged, so it is known as photo-electronic method or photo-electric effect.

In India bids for almost all renewable projects in the recent past have been oversubscribed and the competition has brought down the generation cost. The bids for the first set of solar power projects, announced by the National Democratic Alliance (NDA) government were oversubscribed by ten times. More than thirty developers submitted their bids totalling their 5.5 GW for the 500 MW worth of solar photovoltaic (PV) projects announced under the Jawaharlal Nehru National Solar Mission in Andhra Pradesh. Similarly, in the proposal received for solar projects in Madhya Pradesh and Telengana in July 2025 a developer bid as low as $5.17 per unit of power generated. The benchmark of the Central Electricity Regulatory Commission (CERC), the key regulator of India's power sector, for 2015-16, is $6.86 per unit.

Indian Bids for all Renewable Projects:

i. Rajasthan Solar (514 MW) — 23 Noverber 2025
ii. NHPC Pan-India (1200 MW) — 23 January 2025
iii. NTPC Gujarat (1200 MW) — 14 January 2025
iv. SJVN Pan-India (1500 MW) — 24 January 2025

"The plans seem to be high and ambitious, but can be achieved, if policies like buyback rate, power evacuation, speedy approvals, and fiscal incentives are given by the union and state governments", says O. Subrahmanyam, president, technical division of Axis Energy. The company has pledged to generate 5.5 GW of solar and 7 GW of wind power in the country by 2020. Solar cells and modules manufacturers say that the government has to create a conducive environment to achieve the target.

Challenges and Constraints

The following challenges exist in the generation and distribution of solar energy.

1. Development of essential infrastructure such as transmission lines for evaluation of power.
2. Revenue loss for the developers.
3. Difficulties in acquiring land for the installation of solar panels.
4. It is expensive for average households due to the installation of rooftop solar panels.
5. Constraints in securing finance for the solar power projects.

"The main concern that holds back the domestic manufacturers (in solar) is that the price of solar panels being produced in India is much higer than the imported panels", says K.N. Subramanian, CEO, Moser Baer Solar. The company has committed to install 3 GW of renewable energy projects by 2020.

More than 1000 km^2 such a large area of km^2 land is required for setting up 100 GM solar projects by 2022. Manufacturers say it will be a challenge to acquire such a large area of land in such a short time and then develop them.

Concluding Remarks

Solar power will be made to work in India under the following platform situations. The best option would be to build grid-interactive mini-power plants — also funded through a feed-in tariff paid through the National Clean Energy Fund. But these installations cost more, $15-20 per unit of energy produced and, in addition, the purchasing power is low in remote villages. Therefore, the differential will have to be paid partly through generation-based incentive and partly through tariffs collected locally by the developer. But the key is to make a viable opportunity for investment in providing clean energy to the very poor. A similar model should be evolved for rooftop solar as well. Getting this right would be the real game changer.

The Indian government as well as the state governments should look into the matter on a priority basis to have threefold benefits: clean energy, growth of domestic manufacturing, and meeting the energy needs of all. Conservation and utilization of solar power by the mass will bring down the energy crisis in India.

27

Women in Space Science

Prelude

INTERNATIONAL Women's Day, in the previous years, has been observed on 8 March 2018 everywhere in the world. India has also celebrated this day emphasizing on the women power, especially in the domain of space science. There are so many examples of women as pioneers in space exploration in India and other countries. It is said "Behind every successful man, there is a woman". This signifies the equal, competent, yet silent role that women play in a given success story.

The above dictum appears relevant to the field of space science also. In the sixty plus years of space technology and development, it is marked that space technology is predominantly managed by man. Due to many debatable outlooks, women were either hesitant or neglected in the area of space science. Social disparity and other fears might have been the reasons.

Which analysing the matter, it comes to our attention that even at a time when women were not considered in the field of space science, a few enterprising and bold women played a pivotal role in bringing about a change to this outlook. Thanks to their silent efforts, the scenario is fast changing now.

In the early days of human space flight, some women were engaged in the experimental studies pertaining to the chosen field of aviation and space technology. Their task was to prove that women were equally capable of space travel and the limitations existed only in the mind of the system. A former Soviet cosmonaut Valentina Tereshkova's successful trip to the space station facility and back (Vostok 6 in 1963) disproved inhibitions, opening the door for other women to follow.

Women were not considered eligible to become astronauts. A career in space technology was not an offer for women on the ground that they were not competent enough to face the challenges of this science. In spite of such inhibitions, a few female scientists, driven by intrigue and passion for the field, stood up to challenge the disparity meted out to them. Soon they were proving their mettle in experimentations and ground-based research. Holding critical positions as ground-based engineers, consultants, and technical experts, they helped in shaping the progress of space technology. Their capability in technical positions was found to be more reliable than their male counterparts. They played key roles behind the scene.

The astronauts go through several intense physical and psychological tests before boarding their flights. The ground-based women engineers play a vital role in designing schedules, monitoring the activity of the machinery and astronauts, coordinating communication between the various interfaces, and also making path-breaking changes to the existing system.

In course of time, women spearheaded space exploration by their unique contributions. In many remarkable missions, the role of these brave women was indispensable and critical. Let us discuss here about some unsung women pioneers in the space missions.

Eilene Galloway

The ionic personality is Eilene Galloway (1906–2009), called the "Matriarch of Space Law" and affectionately as "the Grand Dame of Space".

Around 1957, Galloway was working as a congressional researcher, National Defense Analyst and later Senior Specialist in International Relations (National Defense) with the Legislative Reference Service of the Library of Congress in the US. She was in-charge of Writing House and Senate Decuments including those of military importance.

After launching of the first satellite Sputnik by Russia in 1958, US Senator Lyndon B. Johnson proposed the formation of the National Aeronautic and Space Agency (NASA) as a countermeasure in this regard, he sought the help and expertise of Galloway to investigate and help in framing the necessary document to pass the bill. Further, she put forth a proposal to the senator recommending that NASA should be an "Administrative" body with international authority rather than a military "Agency". Feeling the impact of worldwide peace in her recommendation, Johnson immediately made amendments to the Bill and reformed NASA. Soon after, Galloway helped to create the United Nations Committee on the Peaceful Uses of Outer Space (COPUOS).

African-American Women

During the 1940s and 1950s, breaking many social barriers, a group of three African-American women, viz. Katherine Johnson, Dorothy Vaughan, and Mary Jackson, joined the Space Task Force to form a computer pool, initially, recruited on an experimental basis. However, their prowess proved them to be indispensable in their field. Soon the engineers depended on them.

Evidently, they played a crucial role in the Mercury Mission — USA's mission to send the first man into space. Space historians quote that John H. Glenn, the first astronaut to fly on this mission, was wary of the actual calculations made by computers of that time. As the world focused on his historic attempt, the astronaut looked up to Ketherine Johnson. Before finalizing his flight, he insisted that she should run through the computations and check the results. Johnson's girl computer pool ran the calculations, calculated the trajectories and verified the algorithms that the instruments were expected to perform, putting Glenn at ease.

Linda Spilker

Linda Spilker (b. 1955) had been an integral part of the Cassini mission from its inception to the grand finale. Around 1980-81, the Voyager probes passed-by Saturn and beamed back valuable date of the gas giant. During this time, Spilker joined the jet propulsion laboratory as a scientist. The scrolls of date from Voyager were handed over to her group. Mesmerized by it, she poured over the data to study the rings of the planet.

Spilker's idea for sending an exclusive probe to Saturn to study its rings was finally approved by NASA was realized. It was a proud moment for this planetary scientist as she watched the launch and the flawless arrival of Cassini at the gas giant. She was also responsible for designing the famous Cassini's dives through Saturn rings.

At an age of 62, Spilker was busy rooting for another voyage to Enceladus, Saturn's moon which holds the promise of life. Her able assistant and fellow woman scientist Morgan Cable will be taking over from Spilker.

Indian Women in Space Exploration

There are able women scientists at ISRO who play vital roles in its space activities. Women scientists played a prominent role in the Mars Orbiter Mission (MOM). Mangalyaan was a path-breaking achievement for India.

With just eighteen months to realize the project, Anuradha T.K., Senior Programme Director, lead the task force. Ably supporting her were Ritu Karidhal and Nandini Harinath, both Deputy Operation Directors, who had their hands full on the MOM. Also on board were Minal Sampath, Moumita Dutta, and Kriti Faujdar, Systems Engineers.

Within a short time, the team plunged deep into the mission coordinating, conceptualizing, designing, functioning, and troubleshooting the algorithms on the machinery. They provided also solutions to various technical aspects.

Lalithambika V.R., Deputy Director at the Vikram Sarabhai Space Centre (VSSC), Thiruvananthapuram, heads the team that works on fuel optimization for rockets, autopilots, software development for onboard computers, and the hardware to integrate with the equipment. The latest feather in her cap is the smooth launch of 104 Satellites by the PSLV-37. It was a serious task as the satellites' orbits needed careful calculation and each one of them had to be positioned without colliding with each other. These women are currently occupied with Mangalyaan-2 and Chandrayaan-2 missions.

Kalpana Chawla

Kalpana Chawla (17 March 1962 – 1 February 2003) was an Indian-born American astronaut and aerospace engineer who was the first woman of Indian origin to fly to space. Chawla expressed an interest in aerospace engineering from an early age and took engineering classes at Dayal Singh College and Punjab Engineering College in India. She then flied to the USA, where she earned her MSc and PhD degrees becoming a naturalized United States' citizen in the early 1990s.

Sunita Williams

Sunita Williams (born 19 September 1965) is an American astronaut, retired US Navy Officer and one of the most experienced spacewalkers with eight spacewalks and spending a total time of 56 hours and 40 minutes. She was assigned to the International Space Station as a memeber of expedition 14 and 15. In 2012, she served as a flight engineer on expedition 32 and then commander of expediation 33. In 2024, she retured to the ISS of the Boeing Crew Flight Test, the first crewed mission of the Boeing Starliner. Her return to earth has been delayed until March 2025 at the earliest. As such from August 2024 until March 2025, Williams and fellow astronaut Butch Wilmore are pivoting ahead to undertake various scientific experiments and maintenance tasks abroad the ISS.

Epilogue

Above discussions reveal that the women in space science are no less competent than their male counterparts. Obviously, they have set an example for many young girls to enter the field. There are many more instances of such women space scientists of the world who have left their indelible mark on space missions. Their overwhelming contributions in the field have also made them stood in good stead.

28

Physics of the Earth's Atmosphere

Prelude

THE gaseous medium which surrounds the earth is called its atmosphere. With this enveloping atmosphere, the earth continuously rotates about the polar axis while moving along its orbit around the sun, the atmosphere being held bound to the earth by the action of gravity. This gaseous atmosphere is a mechanical mixture of several gases and its composition slightly varies from one locality to another. Besides water vapour, it contains about 77 per cent nitrogen, 21 per cent oxygen, and 1 per cent argon by weight. The remaining 1 per cent includes traces of carbon dioxide, ammonia, hydrogen, neon, krypton, helium, ozone and xenon.

Different Layers of the Atmosphere

In meteorology two types of balloons are used: (i) the recording balloon, and (ii) the pilot balloon. With recent advancements in technology, rockets and artificial satellites are used for conducting investigations of the upper atmosphere.

Recording balloons are hydrogen-filled and automatic recording instruments such as the barograph and thermometers are contained in them. They finally burst out as they ascend higher and higher. The meteorograph, as it lands on the ground, is protected from injury by a special device adopted in the balloon. The heights of the balloons are observed by an instrument called the theodolite.

The gaseous medium constituting the atmosphere cannot expand indefinitely as it extends upwards, for the expansion is finally limited by the action of gravity. The density decreases with the height of the atmosphere increasing, but nothing is definitely known as to the height to which the atmosphere really extends, though it must have a limiting height. Estimates vary. But what is definite is that the atmosphere is more and more rarefied upwards and probably it is finally reduced to a vacuum in the interstellar space. Even the interstellar space, which is ordinarily regarded as the best vacuum, contains traces of matter in the gaseous form, about one atom per cubic antimetre.

Different layers of the atmospheric belt depending on their physical properties are:

1. troposphere,
2. stratosphere,
3. ozonosphere, and
4. ionosphere including the Appleton layer and Heaviside layer up to a height of about 6.5 miles above the earth's surface.

The temperature of the air diminishes steadily as the height increases. The layer of diminishing temperature is known as the troposphere. The layer above this is generally called the stratosphere. Formerly, it was thought that the temperature in this region was constant at about -60° F. However, recent researches reveal that the temperature, in this region increases with height though very slightly. A layer of constant temperature called the tropopause, divides the troposphere from the stratosphere. On the top of the stratosphere an ozone layer has been discovered. This layer absorbs the strong ultraviolet rays coming from the sun and this layer is responsible for reflecting all sound waves travelling upwards from the earth. So, beyond the ozone layer, the region is a valley of silence. It has been confirmed nowadays from studies with radio waves that beyond the ozone layer the upper atmosphere is highly electrically conducting, first pointed out by Balfour Stewart and subsequently insisted on by Arthur Kennelly and Oliver Heaviside and is known as the Kennelly–Heaviside layer. A layer discovered by Appleton above this layer has been named as the Edward Appleton layer.

These conducting layers are only ionized layers of the atmosphere and the existence of such layers has been proved nowadays as a result of intensive radio-investigation since the 1920s. These ionized layers constituting a spherical belt round the earth collectively form the ionosphere. The heights of these layers are liable to both regular and irregular changes during the night or day, and due to various celestial phenomena these are still obscure to us. Radio communication, round the globe the help of short waves, has been possible due to the existence of these conducting layers.

The Atmospheric Pressure

If the whole atmosphere over the surface of the earth is supposed to be divided into a number of layers of air, one above another, then it is evident that the surface of the earth, or any particular layer of air over it, has got to bear the weight of the layers above, and is thus exposed to a pressure which is called the atmospheric pressure. This pressure at a place will, therefore, be equal to the weight of a column of air of unit cross-section and height equal to that of the atmosphere above that place. The value of this pressure is 1,013,961 dynes per sq. cm approximately on the earth's surface and diminishes upwards gradually.

Measurement of the Atmospheric Pressure

The atmospheric pressure is measured by the instruments called barometers. In one type of barometer, the pressure of the atmosphere is measured by the weight of a column of mercury supported by it. Such type of barometers are called mercurial barometers. These are of two types: (i) cistern, and (ii) siphon. Let us illustrate below the construction and working of these barometers.

CISTERN TYPE BAROMETER

This is also called the Fortin's barometer. The barometer tube is filled with pure, dry and air-free mercury and is inverted over a cistern of mercury, *R*, called the resevoir (*fig.* 28.1). The mercury stands in the tube at a certain height. The tube is enclosed within a long brass casing *C* on the front side of the upper part of which there is a rectangular slit through which the upper level of the mercury in the tube can be seen and observed by the help of a small mirror placed on the back side of the tube. The meniscus of the mercury surface is read by a main scale *U*, graduated, in inches and centimetres on the other side of the slit, with the help of a vernier *V*, worked by the knob *P* of a rack and pinion arrangement.

The cistern has its upper part made of a glass cylinder *F* (*fig.* 28.1) through which the surface of the mercury *M* contained in it can be seen. The glass cylinder is fitted in a box-wood cylinder *K* whose lower end is closed by a flexible leather bag *L* (usually made of chamois leather). This bag has a wooden bottom *N* against which the point of the base-screw *S* presses. The screw works through the brass casing *E* which surrounds the reservoir. By turning the base-screw, the level of the mercury in the reservoir can be raised or lowered and finally made to touch the tip of the ivory pointer *I* which is fixed to the lid of the cistern. The tip of the pointer is in the same horizontal level as the zero of the main scale *U* (*fig.* 28.2). The mouth of the cistern is provided with a seat of leather through the pores of which the atmospheric pressure is transmitted from the outside to the inside of the cistern. The barometer tube is of wide bore at the upper portion, so that the effect of surface tension is avoided, but it gradually tapers to a narrow and having a bulge *G* a little higher up, which rests on a leather seat *I* (*fig.* 28.3). The end of the tube is made narrow in order that there may not be oscillations in the mercury during adjustment.

Reading of the Barometer

To read the barometer, the base screw *S* is worked until the mercury surface in the cistern just touches the tip of the ivory pointer. This is called zero-adjustment. The adjustment is accomplished when the tip of the pointer appears to touch the inverted image produced in the clean mercury-surface below. Then keeping the eye in level with the mercury-surface, seen in the slit, the vernier *V* is moved along the main scale *U* until its lower edge (zero-mark) appears to be tangential to the convex mercury-surface, as shown in *fig.* 28.3. The main scale reading below the zero mark of the vernier plus the value of the vernier reading against the main scale gives the barometric height at the time of observation. The temperature of the atmosphere is given by a thermometer *T* placed on the casing of the instrument (*fig.* 28.1).

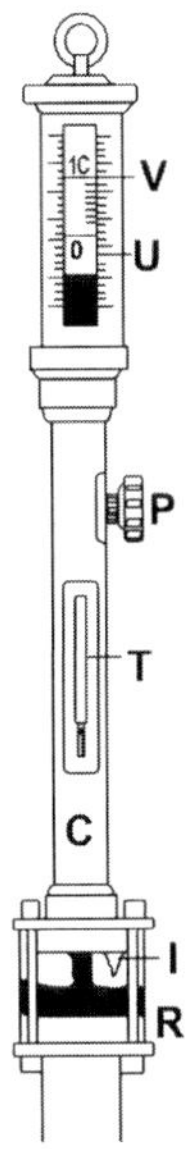

fig.28.1

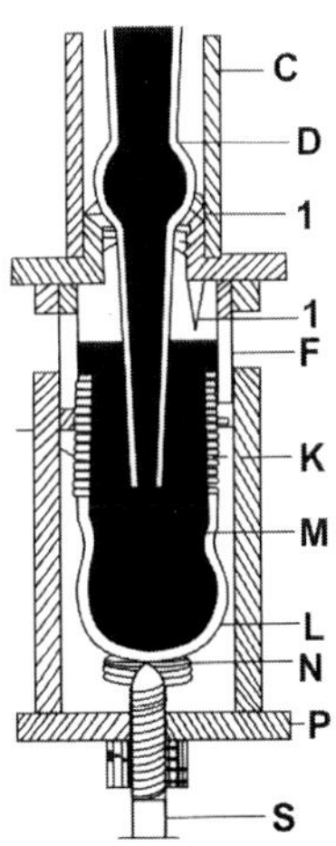

fig.28.2

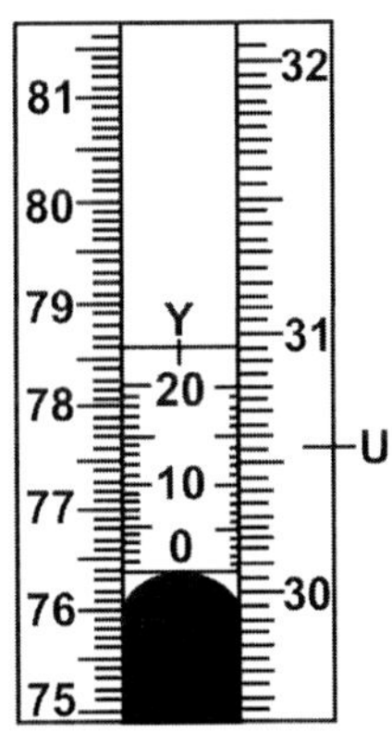

fig.28.3

The Correction of Barometer Readings

The scale reading of a barometer requires to be corrected for temperature, and then certain transformations are necessary so that an observed barometric pressure may be compared with the standard barometric pressure.

1. *Correction for Temperature*: A correction is to be made for the expansion of the metal scale (which is ordinarily supposed to be correct at 0 °C) with rise of temperature. This corrected height in terms of mercury is at the existing temperature. So, this again is to be transformed to zero-degree cold mercury.

2. *Transformation to Sea-level*: As the value of the acceleration due to gravity diminishes with height above the sea-level, transformation is necessary so that the observed reading is reduced to the sea-level.

3. *Transformation to 45° Latitude*: The value of gravity varies from place to place on the surface of the earth. It is less at the equator than at the poles. For correcting the above two effects (2) and (3), the value of gravity at latitude 45° in the sea-level is taken as the standard. The height reduced to sea-level at latitude 45° is given by the relation.

$$H[1 - 0.00257 \cos 2\lambda - 1.06\, s \times 10^{-9}],$$

where, H = observed height reduced to 0° C,

λ = Latitude of the place,

and s = height in cm of the place above the sea-level.

The reading corrected for effects (1), (2), and (3) would represent the height of the mercury column which would be supported by the existing atmospheric pressure at standard temperature (i.e. 0 °C) and at a standard place (i.e. at the sea-level at latitude 45°).

Uses of Barometers

Barometers can be used for the following purposes:

a. Measurement of the atmospheric pressure;
b. forecasting of weather; and

c. determination of the altitude of a place.

SIPHON TYPE BAROMETER

This is a portable type of barometer and is more convenient than the cistern type.

It has no cistern but consists of a *U*-shaped tube having unequal limbs (*fig.* 28.4). The longer limb *AB*, about 1 m long, is closed at the top while the shorter limb *CB* serves as the cistern. To protect the mercury-surface from incidence of impurities, *C* is closed at the top, leaving a small opening at *D* to enable communication with the external air. In order that no air may enter into *AB* when the instrument is inclined, *AB* and *CB* are joined by means of a tube of narrow bore. The instrument is fixed on a wooden board and one scale is fixed on it against each of the two limbs to read the difference in mercury levels, which gives the barometric height at the time of observation.

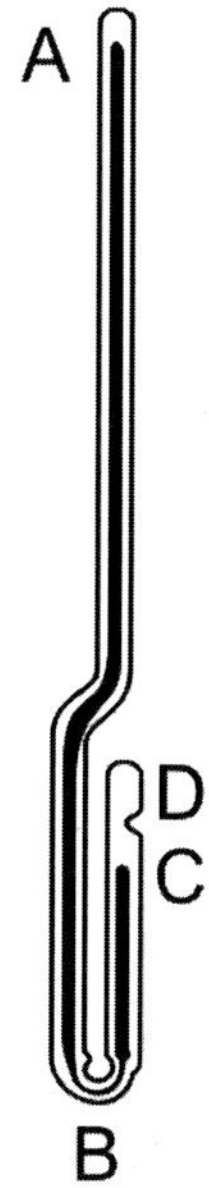

fig. 28.4

29

Outlines of Global Positioning System

Introduction

Global Positioning System (GPS) is a space-based radio navigation system first developed by the US Department of Defence for military use and now available to civilian users throughout the world. Persons using GPS receiver will get information about the location and time. Likewise Global Navigation Satellite System (GLONASS) is being used by the Russians besides the GPS. The European Union has its own version of GPS, known as Galileo, and the Chinese are expanding their local navigation system Bel Dou to a global level.

Components of GPS

A constellation of twenty-four medium earth orbit satellites forms the basis of the GPS. These satellites transmit microwave signals to the GPS receiver. From these signals, the receiver calculates the latitude, longitude, altitude and time for each spacific user. The Russian system has twenty-four satellites in orbit. The proposed EU and Chinese system will be designed basing upon twenty-four and thirty-five satellites respectively.

Basic Principle of GPS

The GPS works on triangulation to find a location. Let us take the conjecture of three straight points. Mathematically three circles can intersect only at one point and hence the position can be traced. Suppose you are at one place which is located at a distance of 115 km from a second place, 185 km from a third place and 262 km from the fourth place. Likewise your position can be traced by intersecting three circles passing through the above-mentioned three places (i.e. second, third, and fourth). Similar principle is used by the GPS to locate the exact positioning of an individual or a place.

Function of GPS

The original GPS design was made up of twenty-seven satellites, out of which twenty-four were operational and three were for backup purposes. Each of these was making two complete rotations at a particular axis around the earth everyday. The satellites were arranged in such a way that from only location on earth at least four, were visible. Since there exist, three variables, viz. latitude, longitude and altitude, three satellites are required for the function. Owing to the high speed motion of the satellites, there is minor time lag between the received signals of the satellites. This causes a large positional error. Therefore, a fourth variable was introduced initially.

Indian Regional Navigation Satellite System (IRNSS)

The Indian Space Research Organisation (ISRO) has launched the sixth navigation satellite of the IRNSS (*fig.* 29.1). This system consists of a constellation of seven satellites. When operation starts, it will provide us our own version of a GPS. It can also be used by mass living in countries that are within 1,500 km of India. IRNSS will provide two types of services, namely a Standard Positioning Service (SPS) which will be available for everyone, and a Restricted Service (RS) which will be provided only to authorized users. Second one is called an encrypted service.

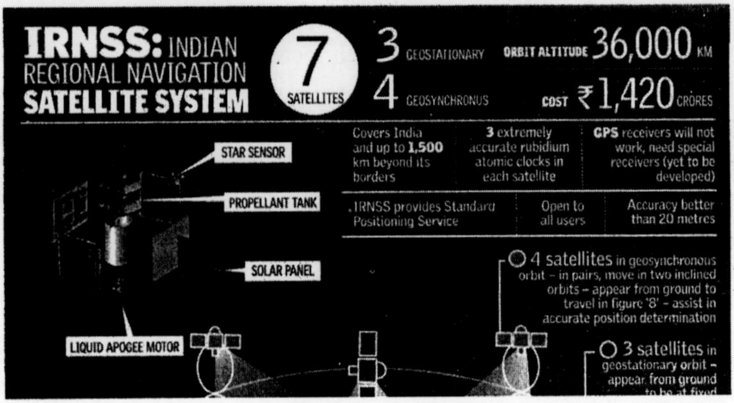

fig. 29.1: Indian Regional Navigation Satellite System (IRNSS)

30

Physics as Conceptualized in Ancient India

Prolegomenon

INDIAN civilization is one of the earliest civilizations of the world. So brilliant and advanced, it left an impact on other civilizations as well, viz. the Egyptian, Persian, and Arabic. From the studies, it is evident that a scientific temperament was existent among the ancient Hindus. They had sound and accurate knowledge not only in the fields of metaphysics and religion, but also in sciences, such as mathematics, astronomy, physics, chemistry, biology, medicine, and engineering. Studies show that they were aware of the systematic streamlined process of investigation of "modern times". The ancient Hindus were not at all strangers to the scientific spirit. They were perfectly familiar with *anusandhāna* or *pariśodhana,* which is nothing but the learned laborious investigation of the modern days. From all available sources of information, the methodology which clearly emerges is analogous to that followed in modern times as that of observation, information, and authoritative expressions.

Logical reasoning and accuracy of observation were the fundamental principles of the Indian scientific age. This process probably had its roots even before the sixth century BCE. The Sāṁkhya and the Vaiśeṣika schools of philosophy had originated and developed this scientific spirit with its emphasis on direct observation, exact measurement, and persistent enquiry. Thus, the scientific thoughts existed and developed in ancient India many centuries before the European Renaissance. The works of scholars, such as Praśastapāda (third-fouth century CE), Āryabhaṭṭa (fifth century CE), Brahmagupta (sixth-seventh century CE), Śabara Svāmī (eighth-ninth century CE), Vācaspati (ninth century CE), Udayana and Śrīdhara Bhaṭṭa (twelfth century CE), Guṇaratna (fourteenth century CE), Dāmodara (sixteenth century CE), and the ancient seats of learning such as Takṣaśilā, Vikramaśilā, Nālandā, Kāñcī and Madurai testify to the continuous intellectual activity in anient India. In fact, great Hindu sages possessed "authentic" scientific temper and never accepted "assumptions" unless theoretical and applied validity was proved.

The present paper deals with some of the ancient scientific ideas on physics latent in the Vedic hymns. Long ago, Vedic bards by their meticulous and continuous observations of various natural events have speculated and theorized many scientific reasonings which owe their relevance to the advancement of knowledge of science in a later period. In this respect, we can say that the Vasiśeṣika Darśana is one of the well-known six systems of the Vedic philosophy.

Contributions of the Indians to Scientific Knowledge

The scientific perceptions of ancient Indian genius are reflected in the concepts of ultimate structure of matter, which was first propounded by the Indians. The evolution of elements which are the building blocks for forming diverse compounds has been discussed in various schools of philosophy in India. Ancient Indians had a fairly good understanding of measuring and mapping the course of heavenly bodies, agricultural techniques, and analysing the constitution of matter. The sources of various scientific perceptions can be traced in the *Ṛgveda*. One is simply wonderstruck to find in certain hymns a searching enquiry into the creation of the world. The song of creation is described in the 129th Sūkta of the 10th Book of the *R̥gveda*. It is called the Nāsadīya Sūkta — the hymn of creation of the universe. Let us throw some light on Vedic cosmogony, i.e. the theory of creation, which is based on physical laws and mechanism.

Vedic Cosmogony

The physics or the science of matter and enery was evolved with the evolution of philosophy of the origin of the existent from the non-existent. Materialism is the oldest known philosophy. It is borne out of human experiences. Then philosophy is evolved as a man's attempt to have a rational explanation of the universe around himself as a part of the universe.

The well-known Puruṣa Sūkta of the *R̥gveda* mentions that everything sprang from the body of a giant personality (*puruṣa*), the Supreme Power. Besides, there are two other cosmogonic hymns in the *R̥gveda*, which explain the origin of the universe as a kind of existent (*sat*) from the non-existent (*asat*). The Nāsadīya Sūkta as translated by A.L. Basam reads as follows:

> Then even, nothingness was not, nor existence. There was no air then, nor the heavens. Beyond it what covered in it? Where was it? In whose keeping?
>
> Was there then cosmic water, in depths unfathomed?
>
> Then there was neither death nor immortality, nor was there then the touch of night and day.
>
> The one breathed windlessly and same sustaining. There was that one then, and there was no other. At first, there was only darkness wrapped in darkness, all this was only unilluminated cosmic water. That one which came to be enclosed in nothing, arose at last born of the power of heat in the beginning desire descended on it.
>
> That was the primal seed born of the mind. The sages who have searched their hearts with wisdom know that which is keen to

> that which is not. And they have stretched their cord across void and know what was above and what was below.
>
> Seminal powers made fertile mighty forces. Below was strength and over it was impulses.
>
> But after all who knows and what can say, whence it all came and how creation heappened?
>
> The *devas* themselves are later than the creation. So who knows truly whence it has arisen?
>
> Whence all creation had its origin, the creator, whether he fashioned it or whether he did not, the creator who surveys it all from highest heaven he knows even he does not know.

Mainly two theories of the creation as the interpretations of the above verse appear in the Vedic and post-Vedic literature which are as follows:

1. Water theory of creation.
2. Shell theory of creation.

Let us discuss each theory in Vedic perspectives.

Water Theory of Creation

According to this theory, only water existed in the beginning. Then evolved the water-borne organisms followed by the entire material world. With the advancement of science and human civilization, this theory became the well-known theory of evolution framed by the famous scientist Charles Darwin.

Shell Theory of Creation

This theory suggests the formation of the material world from a cosmic nucleus, i.e. the golden-egg termed as "Hiraṇyagarbha" in the *Ṛgveda*. The philosophy of creation links this to the Supreme Creator, Prajāpati, who has created the ether, air, fire, water, earth, and finally life. The golden-egg was considered as the source of all matter and energy. All discrete substances of the universe were aggregated here in the golden-egg with enormous energy. The primeval body of this cosmic nucleus was disintegrated by the Superme Spirit (the desire for creation). This was happened with a great explosion evolving sounds of high pitch.

According to the *Chāndogya Upaniṣad,* the universe came to exist after dismemberment of the cosmic egg. This explosion that brought about the creation is nothing but the Indra–Vṛtra collision mentioned in the *Ṛgveda* X.90 and X.121. The modern scienfitic theory calls it the "Big Bang", the great explosion from which the creation has emanated.

Vṛtra, the demon, represents the inertia in physics. Actually inertia opposes the motion. Motion starts overcoming the inertia. Indra threw his *vajra* on Vṛtra and the demon was killed, i.e. inertia was overcome and the motion started. That means the material world began. When Indra hit the demon with his *vajra*, a great explosion occurred evolving huge and tremendous sound. This was later termed as the "Big Bang" in modern science.

The interpretation of the epithet Haima or Hiraṇya (golden) as "bright and reddish yellow" and the word "water" as "water-vapour" paves the way for the existence of a "hot, fireball" at the beginning. The modern cosmological theories have accepted this as the starting point of the material world. This epithet of Haima or Hiraṇya is latent in all the Purāṇas.

In addition to the above two theories of creation, there are some other speculations prevalent in the cosmological hypothesis.

Cosmic Dust Hypothesis

The *Ṛgveda* also refers to the existence of cosmic dust as the building material of the universe. The concerned hymns are addressed to Viśvakarmā who has designed this universe by blasting and smelting.

Creation and Dissolution

The philosophy of the Purāṇas adheres to the notion of the cosmic cycle, i.e. creation, preservation, and dissolution. At the end of the definite period (*kalpas* of the Purāṇas), the material would dissolve and then recreation starts. In the present concept of modern science, matter can be annihilated to become energy in accordance to the famous mass–energy relation of Einstein, i.e. $E = mc^2$. Thus, mass can be transformed into energy and vice versa. According to the Purāṇas, Brahmā creates, Viṣṇu preserves, and Śiva destroys.

The Law of Universal Gravitation

The law of universal gravitation is gleaned from the *ṛk* of the Ṛgvedic hymn. The cosmic order *ṛta* governs the entire universe. According to Sindhughatta Ramasastry Narasimha Murthy, *ṛta* is the law of the Universe or the great cosmic order by which the whole of the manifest universe is working. This, in other words, appears to be the law of gravitation in simple terms. Vedic God Varuṇa is regarded as the lord of this law. Griffiths holds that the sun, the earth, the moon, the air, the water, and the wind are held in their position by this gravitational attraction, and so also all the celestial bodies. The Vedic concept of the attraction of celestial bodies, though brought out by Bhaskara-II about the twelfth century CE did not pave the way for the formulation of the theory of gravitational attraction in India. That was left to Johannes Kepler, Galileo Galilei, Nico laus, Copernicus and Isaac Newton in Europe.

Unified Field Theory

A number of geophysical agencies are traced in the *Ṛgveda* by Murthy. According to him, in the *Ṛgveda* Indra stands for the natural geophysical phenomena. Indra represents the geomagnetic field that brings about earthquakes. Agni stands for geothermal field (heat energy), Varuṇa for the universal law of gravitation, Māruta for the wind-energy, and so on. Thus, there exist a number of forces which govern the universe. These are electrical, magnetic, gravitational, and nuclear forces. The Vedic seers observed that all these forces are the different aspects of one and the same fundamental force. This Ultimate Reality is arrived at and is identified with *Brahman*. Many years after, modern scientists also thought of unifying all these forces which govern the universe. Going by today's physics, four fundamental forces governing the universe are:

1. gravitational force,
2. electromagnetic force,
3. weak nuclear force, and
4. strong nuclear force.

Nobel physicist Abdul Salam has become successful in unifying the electromagnetic force and weak nuclear force. Also Steven Weinberg, independently, has succeeded in unifying the above two forces. Salam predicted his unified theory in 1968, but got Nobel Prize in 1979 along with Sheldon L. Glashow and Steven Weinberg.

Thus, the Salam–Weinberg theory reduced the four fundamental forces into three, viz. the strong nuclear force, the electro-weak force, and the gravitational force. Salam believes that a further extension of the gauge principle is likely to relate the strong interaction to the "electro-weak". Gravity will be taken up to show that all the four forces of nature have a common origin, the ultimate goal of a physicist which was also a speculation made long before by the Vedic seers regarding the "unit source" of all creation.

Space–Time Continuum

In the Vedas, the word *antarikṣa* means the space extending outward from the earth. The Vedic bards considered the duo of the earth and sky when they spoke of *dyāvāpr̥thivī* in many *sūktas*. The earth and the sky are connected by space having air and atmosphere. According to the *Taittirīya Upaniṣad*, it is characterized by the quality of sound. It is noted to give rise to the earth through *vāyu*, *agni*, and *āpas*.

The word *kāla* in the Vedic literature generally means time. The general division of time like *bhūta* (past), *bhavat* (present), and *bhaviṣyat* (future) exists in the Vedas. Other divisions of time like *ahan, māsa, pakṣa, saṁvatsara* or *parivatsara, muhūrtta* and a larger time interval *yuga* are also mentioned in the Vedic literature.

The concept of space and time is significant in astronomy and astrology. There is relative importance of space and time in Indian philosophy. Space and time are taken as separate entities by the Nyāya–Vaiśeṣika school. Both are eternal and all-pervasive. They are coexisting with the entire matter, all events and actions. Space, time, and matter all have sprung from one cosmic and primordial source called *prakr̥ti.* The other view predicts two levels of space and time, viz. *akhaṇḍa* and *sakhaṇḍa. Akhaṇḍa* is one continuous block of time and space, while *sakhaṇḍa* is the divisible space and time. The first represents the nature of *prakr̥ti* and the second symbolizes *ākāśa* under certain conditions. As illustrated by Murthy, each moment is all-pervasive but not eternal. Every moment perishes to give room for the next moment. Time is envisaged as a close associate of continuous evolution of matter and is known intuitively. The reflection of *Brahman* is the space.

Modern science defines space as a medium for light to propagate and that bend near celestial objects. The concept of space as gleaned from the theory of relativity of Einstein is four-dimensional (x, y, z, t) where time has been taken as one of the dimensions, i.e. space is dependent on time. Hence, the space–time continuum is the most fundamental concept that holds this modern scientific thoughts and explanations. This is the object that generates and destroys the material or visible universe.

Vedic Way of Calculating the Speed of Light

In order to calculate the speed of light by the units of distance and time used by the Vedic people, we will follow Sāyaṇācārya's interpretations of the Vedic hymns. Sāyaṇācārya was a minister in the court of Bukka of the great Vijayanagar Empire in south India in early fourteenth century. He comments in his R̥gvedic commentary on a particular verse:

taranir viśvadarśate jyotiṣkradasi sūrya, viśvamā bhśirocanam tathā ca smaryata yojanānām sahasre dve dev śate dve ca yojane ekena nimiṣārdhena kramaman ।

Swift and all beautiful thou, Oh Sūrya! maker of the light, illuminating all the radiant realm traversing 2,202 *yojanas* in half a *nimeṣa*.

Referring to this, Raghava Rao says that Sāyaṇa gives us a concrete idea about the speed of light which travels in half *nimeṣa* 2,202 *yojanas*. One *nimeṣa* is equal to $16/75^{th}$ of a second. Hence, the speed of light can be calculated as follows:

In 1/2 *nimeṣa*, i.e. 8/75 second, light travels 2,202 *yojanas*.

1 *yojana* = 4 *kosas*
1 *kosa* = 2,000 *daṇḍas* (fathoms)
1 *daṇḍa* = 2 yards
1 *yojana* = 2 × 2,000 × 4 yards
= 16,000/1,760 miles
= 100/11 miles (∵ 1 mile = 1,760 yards)

2,202 *yojanas* = 100/11 × 2,202 miles, travelled by light in 8/75 seconds.

So in 1 second, light travels:

100/11 × 2,202 × 75/8 = 1,87,670 miles.

In this manner, the speed of light comes as 1,87,670 miles/second, the modern value being 1,86,000 miles/second. In fact, in 1887, Albert Michelson and Edward Morley conducted experiments to determine the absolute speed of the earth through the hypothetical "either" medium which helped in determining the speed of light.

Conclusion

To sum up, we can say that the speculations of the Vedic seers regarding the origin of the universe, the physical phenomena occurring in the nature, and the universal law governing the universe, all are quite relevant to the modern concepts developed in today's science.

Ostensibly, the science of matter and energy depicted in the Vedas has relevance to our present-day science adhering to some crude notions of the Vedic seers. From the Big Bang to the Big Vision, the Vedic literature is the storehouse of such concepts and speculations which explain scientific terms formulated long after; starting from Newton's law of gravitation to Einstein's mass–energy relation, the Vedas present pre-scientific knowledge, which, later on, is defined as "science". The magico-religious phenomena described in the Vedic literature gave rise to the growth of ancient Indian science, which forms the root of all development in science, in true sense.

31

Sustenance of Life with Light

Prolegomenon

Life cannot be sustained on earth without light. A life needs food, water, and air for its existence. From microorganisms to any living being, these three elements are barely needed for maintenance of life. Nature provides all these to them. To lit the life, man requires entertainment, and it is possible for electromagnetic radiation (light). Electromagnetic radiations available on earth are visible light, ultraviolet light, infrared, microwave, radio wave, X-ray, and gamma ray. Let us throw light on the complete electromagnetic spectrum, which consists of the following waves:

i. Electric waves
   a. Long waves,
   b. standerd broadcasting waves (radio waves), and
   c. microwaves.

ii. Optical region
   a. infrared,
   b. visible light, and
   c. ultra-violet light.

iii. X-rays, and

iv. gamma rays.

Out of the above electromagnetic waves, visible light helps us see an object. So, we define light as a form of energy which produces sensation in sight. Different characteristics of light are:

1. Light is a form of energy.
2. Light is dualistic in nature, i.e. light travels in straight lines as well as in wave form.
3. Light travels with speed 3 × 108 m/sec in vacuum.
4. The speed of light decreases in a material medium.
5. Light is transverse in nature.
6. Light can effect a photographic plate.
7. Light is electromagnetic in nature.

Different phenomena of light are reflection, refraction, interference, diffraction and polarization. Four theories of light have been forwarded by Isaac Newton, Chaistiaan Huygens, James C. Maxwell and Max Planck. These are:

a. Corpuscular theory of light by Newton,
b. Wave theory of light by Huygens,
c. Electromagnetic theory of light by Maxwell, and
d. Quantum theory of light by Planck.

With the above basic knowledge about light, let us discuss how light helps life to exist on the earth. Directly or indirectly light has a vital role to play in the lives of living beings.

Role of Light on the Formation of Food

Food is required by living organisms for supplying energy to lead their life. In order to carry out the day-to-day work, energy is essential. Food is therefore needed to build the body, to repair the wear and tear, and to act as a fuel to supply energy. For preparing food (carbohydrates), plants use sunlight. Photosythesis is the basic chemical process by which plants produce carbohydrates. The green pigment of the levels of plants, the chlorophyll, absorbs energy from the sun and the plants absorb carbon dioxide from the atmosphere, draw water from the soil through the roots. With the help of carbon dioxide, water and sunlight, the plants prepare carbohydrates. The hydrocarbons formed during the process contain mostly sugar and starch. It has been noticed that some plants respond to solar radiation more easily than others. This has led to the suggestion of growing hybrid plants which convert solar to chemical energy so efficiently that they could become a plentiful source of fuel used by man as conventional energy.

Carbohydrates prepared by plants with the help of sunlight are widely distributed throughout the animal and vegetable kingdoms, but are especially found in latter. They include sugar, starch, and cellulose that are present in such substances as rice, wheat, bread, potatoes, fruits and vegetables. A normal adult man eats between 300-500 carbohydrates a day in the form of sugars and starches.

Oxygen for Respiration

During the process of photosynthesis, the plants take carbon dioxide and leave oxygen which is needed for the survival of living beings. Sunshine used in the photosynthesis plays a vital role in the release of oxygen by the plants. Oxygen is the life. Without oxygen no combustion or oxidation can take place and nearly all chemical changes or metabolic processes going on in our body would stop.

Role of Sunlight in the Formation of Water

Water is essential for living beings to sustain their lives. Plants, animals, birds, and all other living organisms require water. They grow and live on food and water. In the case of human beings, water accounts for two-third of the weight of the body. It forms the large part (about 60 per cent) of the tissues of the body and is most essenital requirement for well-being. But, light (sunshine) plays a vital role in the formation of water-cycle. Sunlight falling on seas, rivers, wells, ponds, and other water-reservoirs vapourizes water. Water-vapours go up and form the clouds in the sky. When such clouds are cooled, rains occur, i.e. again water comes to earth and fill up the reservoirs. Thus, the water-cycle is formed in the atmosphere due to sunlight.

Role of Light in Vision

Light plays a major role in the vision of the living beings. We observe the objects around us. Light falling on the lenses of our eyes forms the image of the object on the retina and which is communicated to the brain by the optic nerve that connects the retina and the brain. Then the person sees the object. Eyes are the most precious part of the living creature.

Besides the above, light removes the darkness. White light is a compound light having seven colours or seven wavelengths. These colours are violet, indigo, blue, green, yellow, orange, and red. In short, it is VIBGYOR. A monochromatic light has single colour and hence a single wavelength. When a compound light passes through a prism, it produces a spectrum of its constituent colours. Likewise, the white light spectrum has seven colours, red to violet.

Due to atmospheric refraction of sunlight, stars twinkle. Likewise rainbow is formed because of refraction of sunlight. The rainbow is a beautiful solar spectrum formed when the sunshines on raindrops of fountain spray. The result is due to reflection, refraction, and dispersion of the rays by raindrops, and an observer standing with his back to the sun sees a circular arc of spectral colours in the sky, the red being at the top or the outer edge, and violet at the bottom or the inner edge. This is called a primary rainbow. Sometimes another arc is seen outside the first with the order of the colours reversed, i.e. violet on the top and red at the bottom. This is known as a secondary rainbow.

Miracles of Light Phenomena

Many interesting happenings occur due to reflection and refraction of light. Through a plane mirror, one can see one's image due to reflection of light. Lateral inversion of an image formed in a plane mirror is observed. A better *P* appears *q* in a plane mirror. When an object is in a denser medium and viewed from a rarer medium, the object appears nearer to the observer, i.e. real depth is more than the apparent depth. But, when the object is in a rarer medium and viewed from a denser medium, the object appears to be raised, i.e. the image of the object appears to be farther than the object. Such situations occur due to refraction of light.

A coin placed at the bottom of a basin filled with water appears to be raised from its real position. This bent due to refraction of light. Similarly, a straight rod immersed in water obliquely appears about when viewed above the surface of water. If the rod is immersed vertically in the water, then the part of it under water will appear smaller by one-quarter of its actual length. So, any object will appear to be raised more when viewed more and more obliquely. Standing at one side of a swimming bath, the water directly under you will appear to you to be shallower than it really is, and the water at the far side will appear to be still more shallow, i.e. the bottom more raised up. All these happen due to the phenomenon of refraction of light.

Coloured Light

Light of different colours gives us pleasure. When we observe a fountain in a park dazzling with magnificent colours, it gives us pleasure. Coloured pictures viewed anywhere, either on the curtain of the moving pictures or on the TV screen, provide us mental relaxation.

Street lights are generally neon or mercury. Mercury light seems more calm and soothing than the neon light. Similarly, moon-lit light is very much enjoyable like the mercury light; being a compound light, it consists of thirteen colours. Very recently, Light Emitting Diode (LED) light has come of use. First, LED white light was invented. But, now LED blue light has been invented. In 2014, three scientists got the Nobel Prize in Physics for the invention of LED light. They are Japan's Isamu Akasaki and Hiroshi Amano, and American secientist Shuji Nakamura. These three researchers are selected for their invention of blue light emitting diode. Blue LED light provides power-saving and brilliant light. This LED light is enlightening the screens of our mobile phones and that of our computers. This has developed the standard of more than 150 crore of people. After the invention of LED, low power consumption and brilliant light-emitting bulbs are available at affordable cost.

Royal Swedish Academy of Sciences has said that the invention of LED is only twenty years old. But this has already produced white light in a completely new process and that is employed for our benefit.

It has been declared by the Nobel Committee that these three scientists, in 1990, created the blue light from semiconductor and has transferred the light technique. Using blue light, the white light emitting lamp has been build by the new technique. Further, the committee has also said that the one-fourth use of electricity throughout the world is obtained from the LED sources. Obviously, LED has become capable of resisting the misuse of electricity in a large scale.

Conclusions

Above discussions reveal that light, in whatever form it may be, has become the part and parcel of our life. The entire world of living creatures are benefited by light (sunshine, moonbeam, or of artificial lamps). Light and life go together and these two are inseparable. Proper use of light is beneficial to our lives. Just like water, light should not be wasted. Saving light by any manner paves the way for getting benefit from it at any time. So the slogan of the time is.

> Lit the candle for world peace and welfare of living beings, so that your life will be enlightened with knowledge, happiness, and pleasure.

According to *R̥gveda* I.124.4, dawn is the harbinger of light.

32

Indian Remote-sensing Satellites

INDIAN Space Research Organisation (ISRO) employed its satellite, Bhaskara-I, for remote sensing. The 444-kg experimental satellite was launched by a Russian inter-cosmos rocket in 1979. The near-circular orbit had an apogee of 357 km and an perigee of 512 km. The satellite was spin-stabilized and the spin axis was kept perpendicular to the plane of the orbit. An identical spacecraft, named ", was launched in 1981, again by a Russian rocket. The TV cameras operated in the visible (0.55-0.65 micrometre) and in near-infrared (0.8 micrometre) frequencies. The imagery covered an area of 341 × 341 km with a typical resolution of 1 km. There was also a microwave radiometer on board to register the surface temperature.

The data from Bhaskara-I microwave radiometer were used mainly in land applications and for studying meteorological and oceanographic phenomena. Colour-coding techniques were developed to distinguish different features, such as agricultural land, tropical moist deciduous forest, and the Himalayan dry-temperature forests, besides snow. Several image-processing techniques were also developed.

The TV camera on board Bhaskara-I was not able to function for about five months, because of a technical problem. The camera was later revived. It worked well until the mission was formally terminated with the launch of Bhaskara-II. The microwave radiometer of Bhaskara-II had an additional frequency at 31 GHz, besides the bands carried by the first satellite (19 and 21 GHz). The improved device was able to delineate liquid water and water vapour content in the atmosphere with better accuracy. But the TV camera had to be switched off due to the defect of a component after the first two months.

An improved camera incorporating some of the latest developments in solid-state technology was launched in April 1983 by the satellite launch vehicle, SLV-3, in its second development flight. A snag which developed after a few days in the spacecraft system affecting the camera was eventually corrected. The satellite functioned for nearly seventeen months, exceeding its designed lifetime.

Meanwhile, the National Remote Sensing Agency (NRSA) was set up in Hyderabad to coordinate the activities involving a wide range of users of data. With a view to receiving data from Indian and foreign satellites, a ground station was also established.

Following the experimental missions of Bhaskara-I and Bhaskara-II, ISRO designed and developed an operational remote-sensing satellite, the first of a series under the National (Natural) Resources Management System. A Russian rocket from Baikanur launched the first Indian Remote-sensing Satellite (IRS-1A) on 17 March 1988, into a sunsynchronous orbit of 904 km. The satellite completed one cycle of coverage of the Indian subcontinent in tweenty-two days (i.e. 307 orbits).

The imaging system has three Linear Imaging Self-scanning Sensors (LISS), each with 2048 charge-coupled devices (CCDs). They register the image straight by maximizing the exposure time for each ground point without depending on an oscillating mirror to direct the light as was done by LANDSATs. Each array will scan the line as a push-broom works and successive lines of images are produced by the satellite's movement. Each of the 2,048 elements scan 72.5 m (ground or spatial resolution), which results in a ground swath of 148 km. In an ingenious method, two sensors (LISS-IIA and LISS-IIB) are so placed to get a better resolution, each scanning about 74 km of the swath. With the same number of CCD elements (2,048), the resolution that can be obtained in a 74-km swath is 36.25 m. This way, both LISS-IIA and LISS-IIB together cover the 145-km swath (allowing 3 km for overlap) with a better (about 63 m) resolution.

The four bands of IRS-1A (three in the visible and one in the infrared) have been chosen to maximize the application areas most useful under Indian conditions. Band 1 (in the spectral range of 0.45-0.52 micrometre is useful for coastal environment studies, soil vegetation differentiation and discrimination of coniferous/ deciduous vegetation. Band 2 (0.52-0.59 micrometre) is designed for studying vegetation vigour, rock/soil discrimination, turbidity and bathymetry in shallow waters. Band 3 (0.62-0.68 micrometre) can be used for identifying plant species, while band 4 (0.77-0.86 micrometre) is used for delineation of water features and landforms or gromorphic studies.

The data are converted into a digital format ("bits" that can be decoded by a computer). Electrical signals corresponding to the bits are transmitted. NRSA has the capability to have a quick-look display of any selected spectral-band data and record them on digital tapes.

IRS-1C: First in the Second Generation

The first of the next generation satellites, IRS-1C has severed novel features. It has a high-resolution camera to take panchromatic (PAN) images (black and white in the band, 0.5-0.75 micrometre) with a spatial resolution of less than 10 m. Moreover, the camera can be tilted to view the ground at an angle and produce stereoscopic imagery. The PAN camera is steerable up to 26° across the track, as it can sweep a 70-m wide swath. It can take pictures of the same area every five days. The high resolution is a major step forward and is useful in a wide range of applications, from cartography to district-level planning.

The other camera on board IRC-1C is LISS-3 and has two components — one operating in the visible and near-infrared, and the other in the middle infrared. The visible and near-infrared bands are identical to those in IRS-1A and TRS-1B (in green, red and near-infrared). The middle-infrared band (1.55-1.7 micrometre) is new. It is useful for studying the moisture content in crops and for estimating their yield.

IRS-1C is designed to handle high data rates: 80 and 40 mega bits per second. The spacecraft's main platform houses power, telemetry, telecommand, altitude and orbit control and data-handling systems. All the three cameras, earth sensors and star sensors are mounted on a playload platform, which is thermally separated from the rest of the satellite.

The satellite also carries a Wide-Field Sensor (Wi-FS) that operates over a wide swath of 774 km providing low resolution of 189 m. All three cameras (PAN, LISS-3, and Wi-FS) operate in the push-broom scanning mode in which the images are taken as the satellite sweeps over the track. IRS-1C, launched in December 1995, completed ten years of successful operations in 2005. It orbited the earth nearly 60,000 times and transmitted six lakh imageries.

IRS-P_3

IRS-P_3, launched in 1996, carried two remote-sensing payloads and an X-ray detector. One of them is called Modular Opto-Electronic Scanner (MOS), developed by the German Space Agency. The other instrument is known as Wi-FS, which is similar to that of IRS-1C, but with an additional short-wave infrared band (1.55-1.7 micrometre), sensitive to the moisture content of crops. The band is also useful in differentiating snow and cloud cover and in estimating flood damage. The swath of Wi-FS in TRS-P_3 (810 km) is wider than its counterpart in IRS-1C.

IRS-1D

IRS-1D, which prompted pessimistic forecasts following its initially lower orbit, performed flawlessly. It was launched in September 1997 from India by the first operational launch of a PSLV. Following the corrections carried out by ISRO scientists, the orbit was raised to an acceptable level and excellent imagery was received not only in India but also in the US, Germany, Japan, Dubai, Korea and Thailand. The cameras were operated 10-14 times every day to meet the demands.

Oceansat

IRS-P_4 was launched in May 1999. It is prepared by the technology of India. It is the first in a series of ocean satellites planned by India. The satellite has two payloads: Ocean Colour Monitor (OCM) and a Multi-frequency Scanning Microwave Radiometer (MSMR). OCM, a solid-state camera, operates in eight-narrow channels in the visible and near-infrared region of the spectrum. OCM has a swath of 1,420 km. Its data are useful in identifying potential fishing zones in coastal waters, exploration of deep-sea fishery resources, and in the prediction of shoreline changes, circulation and dispersal pattern, sediment movement marine pollution and in coral reef studies. The data would also be useful for many applications on land, such as estimating water run-off.

MSMR operates in four microwave frequencies and can penetrate clouds. From the data it collects, it can derive geophysical parameters, such as ocean surface winds, sea surface temperature, cloud water content and atmospheric water vapour content above the ocean. The combination of optical and microwave sensors monitoring the same area concurrently help in the study of currents. MSMR has a swath of 1,360 km. The satellite at 720 km can see the same area once in two days. MSMR has been configured in such way that it would have the same angle as it scans the area within its footprints.

A Technology Experiment Satellite was launched in 2001 to validate and demonstrate technologies for cartographic applications. The spatial resolution of the imagery taken was one metre. The satellite also tried improvised reaction wheels with a single propellant tank, and altitude and orbit-control systems.

Resourcesat-1

In 2003, Resourcesat-1 was launched. It is the most advanced remote-sensing satellite of ISRO. A PSLV rocket placed it in 820-km polar sunsynchronous orbit. It is a replacement of IRS-1C and IRS-1D. It has advanced sensors for land and ocean resource applications.

Cartosats

Cartosat-1 was India's first satellite for cartography. In 2005, it was launched successfully from Sriharikota. The satellite is designed to make a difference in mapping the country's natural resources. It is rock-steady in its orbit with two GPS receivers on board.

As the satellite traces a circular orbit around the earth at a height of 617 km, it is designed to revisit the same area every five days. Cartosat's imaging technique is quite sophisticated. The satellite has earth sensors, yaw sensors, a wide area star sensor and precision digital sun sensors to determine the altitude in orbit.

Cartosat-2 was launched in 2007. It has a single PAN camera that can take scene-specific images at a spatial resolution of better than 1 m covering a swath of 12 km.

Low-Earth Orbit (LEO) Satellites

The ISRO Telemetry, Tracking and Command Network (ISTRAC) supports Indian satellites in low-earth orbit and launch vehicle missions. Radar satellites have also been developed by ISRO. Apart from crop estimate, disaster management, especially during floods, is expected to be more timely and effective, based on the radar imagery.

Applications

Space technology and advances in electronic sensors have improved the prospect of using the natural resources in a sustainable manner. Remote-sensing data obtained and processed from satellite-based sensors have found numerous applications in the management of natural resources.

33

Exploring Mars

LANDING on the moon by man has paved the way for exploring the other planets close to the earth. Though Mercury and Venus are nearer to the earth, the choice for man-landing has fallen on Mars mainly due to the surface and atmospheric conditions prevalent on this planet as observed through earth-based instruments and space probes. There are also speculations about the possibility of tracing some form of "life" on Mars. These speculations are primarily based on some almost identical, physical conditions prevailing on Mars and the earth.

Geometry

Mars is almost equal to the earth in size. Its diameter is about 6,800 km, while that of the earth is about 6,380 km. The tilt of the axis of rotation of Mars and that of the earth are also the same, about 23.5°.

Gravity

The value of acceleration due to gravity (g) on Mars is low. It is only 775 cm/sec^2, i.e. about one-third that on the earth. A Mars Excursion Module landing on the planet would have to overcome much less gravitational pull to return to the earth. The ecaspe velocity would be only about 3.1 km/sec compared to 11.2 km/sec on the earth.

Temperature

The day temperatures on Mars are around 30 °C, comparatively the same as on the earth. The night temperatures on Mars go down to about -120 °C.

Duration of "Day" and "Night"

The duration of Martian day and night is of 24 hours which is the same as on our planet. This is the most striking feature of Mars.

Seasons

The Martian seasons, though of different durations, are also like ours. The winter on Mars lasts for 160 earth-days, spring for 199, summer for 182 and autumn 146.

Atmosphere of Mars

The atmosphere of Mars is different from that of the earth. The Martian atmosphere is comparatively thinner. It was believed to contain nitrogen (94 per cent), water vapour, and very little oxygen (less than 0.1 per cent). Mariner-4 sent by USA in July 1965 could not detect any other gas except carbon dioxide and water vapour although it passed within 13,000 km of the planet. Absence of adequate oxygen in the Martian atmosphere does exclude the possibility of existence of "life" of the form known on earth, but it does not exclude completely the existence of some other "life" if any, from deadly solar and cosmic radiations. Red colour of Mars suggests that there may be some minerals like iron oxide on the planet. The presence of iron oxide provides the required protection by absorbing the radiations.

The data collected by Mariner-6 and Mariner-7 sent by USA to Mars reveal that there is hardly any trace of "life" on Mars. These two space-probes flew past the planet within 3,200 km.

The solar radiation received in the region of Mars is less than half that received near the earth. This, coupled with the fact that the Martian atmosphere is very thin, makes the temperature very low. Despite the low temperature and an atmosphere of carbon dioxide with small amounts of water vapour, Martian conditions are similar to those encountered in the frozen lakes of the Antarctica where microbial life has been detected. In extreme environmental conditions blue green algae survive.

Cloud Formations

Well-known clouds are formed in the Martian atmosphere. These occur due to ice crystals at high altitudes (like in terrestrial clouds). There is also the formation of blue clouds at the equator at rising and setting limbs. Dust storms are rarely caused in the atmosphere of Mars. In 1911, for instance, a vast cloud was observed. This cloud hung over much of the southern hemisphere and persisted for months. Since the winds on Mars are not capable of raising such gaint dust storms, their genesis is yet under investigation. One possibility may be that the planet is very close to the asteroidal belt, which is the source of meteorites, and hence it might be bombarded by huge meteorites that raise such dust storms. Some sort of tidal effects could also be responsible for the dust storms as they form almost exclusively during the perihelion passage.

As the dust does not reach great heights, clear clouds can be noticed in the overlying layers. Reduced brightness in the polar regions is observed by the telescope when the dust storm disappears. Thus, it is obvious that the polar regions are polluted by the dust.

Rainfall

Compared to the terrestrial atmosphere, the Martian atmosphere contains very small amount of water vapour. It is less by a factor of more than 100. Such a scanty amount of water vapour is the cause of no rainfall on Mars.

Volcanic Action and Water-Loss

The analysis of date obtained from Mariner-1 reveals that Mars is losing something like 380,000 litres of water a day from its atmosphere. According to the findings of Mariner project scientists, the water vapour may be from gases vented from deep within the planet by volcanic action. The escaping water vapour with considerably greater amounts of carbon dioxide from the south polar ice cap forms a deck of white clouds high in the Martian atmosphere (i.e. stratosphere) and individual cloud puffs at much lower altitudes.

The most interesting and exciting part of the discovery is that Mars is much more active (geologically) than the moon, but less than the earth. Harold Masursky, team-leader for TV experiments aboard Mariner-9, predicted the existence of volcanoes on Mars. This confirms how the surface of the planet is alive.

The Polar Regions

The polar regions may be formed with water and solid carbon dioxide. During the summer, the polar cap shrinks. As a result, a dark fringe appears along the edge of the cap. This is perhaps developed due to the moistening of the land around the receding ice caps. In the polar regions, local bright spots are also visible there, where ice lasts longer. These spots are caused by the recondensation of vapour at higher elevations from the ground.

The Dark Green Regions

The climatic variations of the dark green regions are mostly observed in the equatorial parts. The most noticeable dark areas are Syrtis Major in the southern hemisphere and Mare Acidalium in the northern hemisphere. When the polar ice melts, these regions show a deepening of colour.

The dark regions grow still darker. In the presence of water in the atmosphere, typical vegetation grows. The climatic variations in the dark regions are attributed to vegetation and hence support the assumptions of living organisms on Mars. The other factor favouring life on Mars is the existence of straight rilles, some of which probably connect the dark regions with the red desert regions.

The Bright Deserts

The bright deserts on the surface of Mars are the prominent regions having the colour of the common terrestrial igneous rock, rhyolite. The dark regions and the red desert regions are connected by some of the rilles mentioned earlier. These rilles were assumed as the straight canals built by the intelligent people of Mars for irrigating the desert fields. Though this theory of rilles was supported by the famous astronomer Percival Lowell, others discard the theory arguing that it was more psychological rather than astronomical.

Conclusion

Above discussions unveil that the surface of Mars is alive. In the polar regions, ice caps are formed during the winter and are melted and shrunk during the summer. The dark regions on the surface show activity with the climatic cycle. These regions change their colour with the liquification of the polar ice. White and blue clouds form in the atmosphere and occasionally dust storms swirl about.

Mars is very cold, without much water, without oxygen, with a very thin atmosphere which lets in lethal ultraviolet rays from the sun. Evidently, the existence of life of any kind on Mars becomes difficult. The Mariners and the Russian Mars probes have given us valuable information about the planet. Future voyagers to Mars will probably settle the controversy over the existence or non-existence of life on Mars.

34

Observing the Planet Venus

Introduction

OBSERVATIONS of the planet Venus include those in antiquity, telescopic notes, and from visiting spacecraft. Spacecraft have performed various fly bys, orbits, and landings on Venus, including balloon probes floated in the atmosphere of Venus. The study of the planet is aided by its relatively close proximity to the earth, compared to other planets, but the surface of Venus is obscured by an atmosphere opaque to visible light. Venus is always brighter than the brightest stars outside the solar system.

Venus is the second planet from the sun and is named after the Roman Goddess of Love and Beauty. As the brightest natural object in earth's night sky after the moon, Venus can cast shadows and can be visible to the naked eye in broad daylight. Venus's orbit is smaller than that of the earth, but its maximal elongation is 47°; thus, at latitudes with a day–night cycle, it is most readily visible for up to a few hours following the start of sunset or before sunrise. At times, it has been seen in a completely dark sky. Venus orbits the sun every 224.7 earth days. It has a synodic day length of 117 earth days and a sidereal rotation period of 243 earth days. Consequently, it takes longer to rotate about its axis than any other planet in the solar system and does so in the opposite direction to all but Uranus.

This means that the sun rises from its western horizon and sets in its east. Venus does not have only moons, a distinction it shares only with Mercury among the planets in the solar system.

Orbital Characteristics

Aphelion	0.728213 AU; 108,939,000 km
Perihelion	0.718440 AU; 107,477,000 km
Semi-major axis	0.723332 AU; 108,208,000 km
Eccentricity	0.006772
Orbital period (sidereal)	224.701 d; 0.615198 yr; 1.92 Venus solar days
Orbital period (synodic)	583.92 days
Average orbital speed	35.02 km/s
Mean anomaly	50.115°
Inclination	3.39458° to ecliptic 3.86° to the sun's equator 2.15° to invariable plane
Longitude of ascending node	76.680°
Argument of perihelion	54.884°
Satellites	None

Physical Characteristics Data

Mean radius	6,051.8 ± 1.0 km
Flattening	0
Surface area	4.6023×10^{8} km^2
Volume	9.2843×10^{11} km^3
Mass	4.8675×10^{24} kg
Mean density	5.243 g/cm^3
Surface gravity	8.87 m/s^2
Escape velocity	10.36 km/s

Synodic rotation period	-116.75 d (retrograde); 1 Venus solar day
Sidereal rotation period	-243.0226 d (retrograde)
Equatorial rotation velocity	6.52 km/h (1.81 m/s)
Axial tilt	2.64° (for retrograde rotation); 177.36° (to orbit)
North Pole right ascension	18h 11m 2s; 272.76°
North Pole declination	67.16°
Albedo	0.689 (geometric); 0.76 (bond)
Mean surface temperature	464 °C

Atmosphere

Surface Pressure	93 bar (9.3 MPa); 92 atm

Composition by Volume

Carbon dioxide	96.5 per cent
Nitrogen	3.5 per cent
Sulphur dioxide	0.015 per cent
Argon	0.0070 per cent
Water vapour	0.0020 per cent
Carbon monoxide	0.0017 per cent
Helium	0.0012 per cent
Neon	0.0007 per cent

Besides, there are traces of carbonyl sulphide, hydrogen chloride and hydrogen fluoride.

Rotation is defined by retrograde, as done by NASA Space missions and the USGS, puts Ishtar Terra in the northern hemisphere and makes the axial tilt 2.64°. Following the right-hand rule for prograde, rotation puts Ishtar Terra in the southern hemisphere and makes the axial tilt 177.36°.

The third smallest planet in the solar system, Venus is a terrestrial planet and is sometimes called earth's "sister planet" because of their similar size, mass, proximity to the sun, and bulk composition. It is radically different from the earth in other respects. It has the densest atmosphere among the four terrestrial planets, consisting of more than 96 per cent carbon dioxide. The atmospheric pressure at the planet's surface is about 92 times the sea level pressure of the earth or roughly the pressure at 900 m (3,000 ft) underwater. Even though Mercury is closer to the sun, Venus has the hottest surface of all planets in the solar system, with a mean temperature of 737 °K (464 °C). Venus is shrouded by an opaque layer of highly reflective clouds of sulphuric acid, preventing its surface from being seen from the earth. It may have had water oceans in the past, but after their evaporation the temperature rose under a runaway greenhouse effect. The water has probably photodissociated and the free hydrogen has been swept into interplanetary space by the solar wind, because of the lack of a planetary magnetic field. Owing to the lethal surface conditions, the planet is sometimes referred to as the earth's "evil twin".

As one of the brightest objects in the sky, Venus has been a major fixture in human culture for as long as records have existed. It has been made sacred to gods of many cultures and has been a prime inspiration for writers and poets as the "morning star" and "evening star". Venus was the first planet to have its motions plotted across the sky, as early as the second millennium BCE.

Its proximity to the earth has made Venus a prime target for early interplanetary exploration. It was the first planet, other than the earth, visited by a spacecraft (Venera-1 in 1961) and the first to be successfully landed on (Venera-7 in 1970). The planet's thick clouds render observation of its surface impossible in the visible spectrum and the first detailed maps did not emerge until the arrival of the Magellan orbiter in 1991. Plans have been proposed for rovers or more complex missions, but they are hindered by Venus's hostile surface conditions. The possibility of life on Venus has long been a topic of speculations; in recent years, the topic has received active research attention.

Physical Characteristics in Version

Venus is one of the four terrestrial planets in the solar system, meaning that it is a rocky body like the earth. It is similar to the earth in size and mass, and is often described as earth's "sister" or "twin" as mentioned earlier. The diameter of Venus is 12,103.6 km – only 638.4 km less than the earth's – and its mass is 81.5 per cent of the mass of the earth. Conditions on the Venusian surface differ radically from those on the earth, because its dense atmosphere is consisted of 96.5 per cent carbon dioxide with most of the remaining 3.5 per cent being nitrogen. The surface pressure is 9.3 megapscals (93 bars) and the average surface temperature is 737 K (464 °C), above the critical points of both major constituents and making the surface atmosphere a supercritical fluid.

Atmosphere and Climate

Venus's atmosphere is extremely rich in primordial noble gases compared to that of the earth. This enrichment indicates an early divergence from the earth in evolution. An unusually large comet impact or accretion of a more massive primary atmosphere from solar nebula has been proposed to explain the enrichment. However, the atmosphere is also depleted of radiogenic argon, a proxy to mantle degassing, suggesting an early shutdown of major magnetism.

Studies have suggested that billions of years ago, Venus's atmosphere could have been much more like the one surrounding the early earth and that there may have been substantial quantities of liquid water on the surface. After a period of 600 million to several billion years, solar forcing from rising luminosity of the sun caused the evaporation of the original water. A runway greenhouse effect was created once a critical level of greenhouse gases (including water) was added to its atmosphere. Although the surface conditions on Venus are no longer hospitable to any earth-like life that may have formed before this event, there is speculation on the possibility that life exists in the upper cloud layers of Venus,

50 km up from the surface, where the temperature ranges between 303 and 353 K (30 and 80 °C), but the environment is acidic. The putative detection of an absorption line of phosphine in Venus's atmosphere, with no known pathway for abiotic production, led to speculation in September 2020 that there could be extant life currently present in the atmosphere. Later researches attributed the spectroscopic signal that was interpreted as phosphine to sulphur dioxide, found that, in fact, there was no absorption line.

Thermal inertia and the transfer of heat by winds in the lower atmosphere mean that the temperature of Venus's surface does not vary significantly between the planet's two hemispheres, those facing and not facing the sun, despite Venus's extermely slow rotation. Winds at the surface are slow, moving at a few kilometres per hour, but because of the high density of the atmosphere at the surface, they exert a significant amount of force against obstructions, and transport dust and small stones across the surface. This alone would make it difficult for a human to walk through, even without the heat, pressure, and lack of oxygen.

Above the dense CO_2 layer, there are thick clouds, consisting mainly of sulphuric acid, which is formed by sulphur dioxide and water through a chemical reaction resulting in sulphuric acid hydrate. Additionally, the atmosphere consists of aprproximately 1 per cent ferric chloride. Other possible constituents of the cloud particles are ferric sulphate, aluminium chloride, and phosphoric anhydride. Clouds at different levels have different compositions and particle size distributions. These clouds reflect and scatter about 90 per cent of the sunlight that falls on them back into space, and prevent visual observation of Venus's surface. The permanent cloud cover means that although Venus is closer to the sun than the earth it receives less sunlight on the ground. Strong 300 km/h (185 mph) winds at the cloud tops go around Venus about every four to five earth days. Winds on Venus move at up to 60 times the speed of its rotation, whereas the earth's fastest winds are only 10-20 per cent of its rotation speed.

The surface of Venus is effectively isothermal; it retains a constant temperature not only between the two hemispheres but between the equator and the poles. Venus's minute axial tilt is less than 3°, compared to 23° on the earth, also minimizes seasonal temperature variation. Altitude is one of the few factors that affect Venusian temperature. The highest point on Venus, Maxwell Montes, is therefore the coolest point, with a temperature of about 655 K (380 °C) and an atmospheric pressure of about 4.5 MPa (45 bar). In 1995, the Magellan spacecraft imaged a highly reflective substance at the tops of the highest mountain peaks that bore a strong resemblance to terrestrial snow. This substance was likely formed from a similar process of snow formation, albeit at a far higher temperature. Too volatile to condense on the surface, it rose in gaseous form to higher elevations, where it is cooler and could precipitate. The identity of this substance is not known with certainty, but speculation has ranged from elemental tellurium to lead sulphide (galena).

Although Venus has no seasons as such, in 2019, astronomers identified a cyclical variation in sunlight absorption by the atmosphere, possibly caused by opaque, absorbing particles suspended in the upper clouds. The variation causes observed changes in the speed of Venus's zonal winds and appears to rise and fall in time with the sun's eleven-year sunspot cycle.

The existence of lightening in the atmosphere of Venus has been controversial, since the first suspected bursts were detected by the Soviet Venera probes. In 2006-07, Venus Express clearly detected whistler mode waves, the signatures of lightening. Their intermittent appearance indicates a pattern associated with weather activity. According to these measurements, the lightening rate is at least half of that on the earth, however, other instruments have not detected lightening at all. The origin of any lightening remains unclear, but they could originate from the clouds or volcanoes.

In 2007, Venus Express discovered that a huge double atmospheric vortex exists at the south pole of Venus. It also discovered, in 2011, that an ozone layer exists high in the atmosphere of Venus. On 29 January 2013, ESA scientists reported that the ionosphere of Venus streams outwards in a manner similar to the ion tail seen streaming from a comet under similar conditions.

In December 2015, and to a lesser extent in April and May 2016, researchers working on Japan's Akatsuki Mission observed bow shapes in the atmosphere of Venus. This was considered as a direct evidence of the existence of perhaps the largest stationary gravity waves in the solar system.

Geography

This section includes the discussions of mapping, geology and surface features of Venus.

The Venusian surface was a subject of speculation until some of its secrets were revealed by planetary science in the twentieth century. Venera landers in 1975 and 1982 returned images of a surface covered with sediments and relatively angular rocks. The surface was mapped in detail by Magellan in 1990-91. The ground shows the evidence of extensive volcanism and the presence of sulphur in the atmosphere may indicate that there have been recent eruptions.

About 80 per cent of the Venusian surface is covered by smooth, volcanic plains, consisting of 70 per cent plains with wrinkle ridges and 10 per cent smooth or lobate plains. Two highland "continents" make up the rest of its surface area, one lying in the planet's northern hemisphere and the other just south of the equator. The northern continent is called Ishtar Terra after Ishtar, the Babylonian Goddess of Love and is about the size of Australia Maxwell Montes, the highest mountain on Venus, lies on Ishtar Terra. Its peak is 11 km above the Venusian average surface elevation. The southern continent is called Aphrodite Terra, named after the Greek Goddess of Love, and is larger of the two highland regions at roughly the size of South America. A network of fractures and faults covers much of this area.

The absence of evidence of lava flow, accompanying any of the visible calderas, remains an enigma. The planet has a few impact craters, demonstrating that the surface is relatively young, at 300-600 million years old. Venus has some unique surface features in addition to the impact craters, mountains, and valleys commonly found on rocky planets. Among these are flat-topped volcanic features called "farra", which look somewhat like pancakes and range in size from 20 to 50 km across and from 100 to 1,000 m (330 to 3,280 ft) in height; radial, star-like fracture systems called "novae"; features with both radial and concentric fractures resembling spider webs, known as "arachnoids"; and "coronae", circular rings of features sometimes surrounded by a depression. These features are volcanic in origin.

Most Venusian surface features are named after historical and mythological women. Exceptions are Maxwell Montes, named after James Clerk Maxwell, and highland regions Alpha Regio, Beta Regio, and Ovda Regio. The last three features were named before the current system was adopted by the International Astronomical Union, the body which oversees planetary nomenclature.

The longitude of physical features on Venus are expressed relative to its prime meridian. The original prime meridian passed through the radar-bright spot at the centre of the oval feature Eve, located south of Alpha Regio. After the Venera missions were completed, the prime meridian was redefined to pass through central peak in the Crater Ariadne on Sedna Planitia.

The stratigraphically oldest tessera terrains have consistently lower thermal emissivity than the surrounding basaltic plains measured by Venus Express and Magellan, indicating a different, possibly a more felsic, mineral assemblage. The mechanism to generate a large amount of felsic crust usually requires the presence of water ocean and plate tectonics, implying that habitable condition had existed on early Venus. However, the nature of tessera terrains is far from certain.

Volcanism

Much of the Venusian surface appears to have been shaped by volcanic activity. Venus has several times as many volcanoes as the earth, and it has 167 large volcanoes that are over 100 km across. The only volcanic complex of this size on the earth is the Big Island of Hawaii. This is not because Venus is more volcanically active than the earth, but because its crust is order and is not subject to the same erosion process. Oceanic crust of the earth is continually recycled by subduction at the boundaries of tectonic plates and has an average age of about 100 million years, whereas the Venusian surface is estimated to be 300-600 million years old.

Several lines of evidence point to the ongoing volcanic activity on Venus. Sulphur dioxide concentrations in the atmosphere dropped by a factor of ten between 1978 and 1986, jumped in 2006, and again declined tenfold. This may mean that levels had been boosted several times by large volcanic eruptions. It has also been suggested that Venusian lightening (noted below) could originate from volcanic activity (i.e. volcanic lightening). In January 2020, astronomers reported evidence suggesting that Venus is currently volcanically active, specifically with the detection of olivine, a volcanic product that would weather quickly on the planet's surface.

In 2008 and 2009, Venus Express observed the first direct evidence of ongoing volcanism in the form of four transient localized infrared hot spots within the rift zone Ganis Chasma, near the shield volcano Maat Mons. Three of the spots were observed in more than one successive orbit. These spots are thought to represent lava freshly released by volcanic eruptions. The actual temperatures are not known, because the size of the hot spots could not be measured, but are likely to have been in the 800-1,100 K (527-827 °C) range, relative to a normal temperature of 740 K (467 °C).

Craters

Almost a thousand impact craters on Venus are evenly distributed across its surface. On other cratered bodies, such as the earth and the moon, craters show a range of states of degradation. On the moon, degradation is caused by subsequent impacts, whereas on the earth it is caused by wind and rain erosion. On Venus, about 85 per cent of the craters are in prestine condition. The number of craters, together with their well-preserved condition, indicates that the planet underwent a global resurfacing event 300-600 million years ago, followed by a decay in volcanism. Whereas earth's crust is in continuous motion, Venus, is thought to be unable to sustain such a process. Without plate tectonics to dissipate heat from its mantle, Venus, instead, undergoes a cyclical process in which mantle temperatures rise until they reach a critical level that weakens the crust. Then, over a period of about 100 million years, subduction occurs on an enormous scale, completely recycling the crust.

Venusian craters range from 3 to 280 km in diameter. No craters are smaller than 3 km, because of the effects of the dense atmosphere on incoming objects. Those objects with less than a certain kinetic energy are slowed so much by the atmosphere that they do not create an impact crater. Incoming projectiles less than 50 m in diameter will fragment and burn up in the atmosphere before reaching the ground.

Internal Structure

Little direct information is available about the internal structure and geochemistry of Venus. The similarity in size and density between Venus and the earth suggests they share a similar internal structure: core, mantle, and crust. Like that of the earth, the Venusian core is most likely at least partially liquid because the two planets have been cooling at about the same rate, although a completely solid core cannot be ruled out. The slightly smaller size of Venus means pressures are 24 per cent lower in its deep interior than that of the earth. The predicted values for the moment of inertia, based on planetary models, suggest a core radius of 2,900–3,450 km. This is in line with the first observation-based estimate of 3,500 km.

The principal difference between the two planets is the lack of evidence for plate tectonics on Venus, possibly because its crust is too strong to subduct without water to make it less viscous. This results in reduced heat loss from the planet, preventing it from cooling and providing a likely explanation for its lack of an internally generated magnetic field. Instead, Venus may lose its internal heat in periodic major resurfacing events.

Magnetic Field and Core

In 1967, Venera-4 found the magnetic field of Venus much weaker than that of the earth. This magnetic field is induced by an interaction between the ionosphere and the solar wind, rather than by an internal dynamo as in the earth's core. It is Venus's small induced magnetosphere against cosmic radiation.

The lack of an intrinsic magnetic field at Venus was surprising, given that it is similar to earth in size and was expected also to contain a dynamo at its core. A dynamo requires three things: a conducting liquid, rotation, and convection. The core is thought to be electrically conductive and, although its rotation is often thought to be too slow, simulations show it as adequate to produce a dynamo. This implies that the dynamo is missing, because of lack of convection in Venus's core. On the earth, convection occurs in the liquid outer layer of the core, because the bottom of the liquid layer is much higher in temperature than the top. On Venus, a global resurfacing event may have shut down plate tectonics and led to a reduced heat flux through the crust. This insulating effect would cause the mantle temperature to increase, thereby reducing the heat flux out of the core. As a result, no internal geodynamo is available to drive a magnetic field. Instead, the heat from the core is reheating the crust.

One possibility is that Venus has no solid inner core, or that its core is not cooling, so that the entire liquid part of the core is approximately at the same temperature. Another possibility is that its core has completely solidified already. The state of the core is highly dependent on the concentration of sulphur, which is unknown of present.

The weak magnetosphere around Venus means that the solar wind is interacting directly with its outer atmosphere. Here, ions of hydrogen and oxygen are being created by the dissociation of water molecules from ultraviolet radiation. The solar wind then supplies energy that gives some of these ions sufficient velocity to escape the gravity field of Venus. This erosion process results in a steady loss of low-mass hydrogen, helium, and oxygen ions, whereas higher-mass molecules are to be retained. Atmospheric erosion by the solar wind could have led Venus to the loss of most of its water during the first billion years after its formation. However, the planet may have retained a dynamo for its first 2-3 billion years, so the water loss may have occurred more recently. The erosion has increased the ratio of higher-mass deuterium to lower-mass hydrogen in the atmosphere 100 times compared to the rest of the solar system.

Cultural Aspects of Venus

Venus is a primary feature of the night sky and so it has been of remarkable importance in mythology, astrology and fiction throughout history and in different cultures.

In Sumerian religion, Inanna was associated with the planet Venus. Several hymns praise Inanna in her role as the Goddess of the planet Venus. Theology professor Jeffrey Cooley has argued that, in many myths, Inanna's movements may correspond with the movements of the planet Venus in the sky. The discontinuous movements of Venus relate to both mythology as well as Inanna's dual nature. Unlike any other deity, Inanna is able to descend into the nether world and return to the heavens. Venus appears to make a similar descent, setting in the west and then rising again in the east. An introductory hymn describes Inanna leaving the heavens and heading for Kur, what could be presumed to be the mountains, replicating the rising and setting of Inanna. In Inanna and Shukaletuda and Inanna's Descent into the nether world appear to parallel the motion of the planet Venus. In Inanna and Shukaletuda, the latter is described as scanning the heavens in search of Inanna, possibly searching the eastern and western horizons. In the same myth, while searching for her attacker, Inanna herself makes several movements that correspond with the movements of Venus in the sky.

Classical poets such as Homer, Sappho, Ovid, and Virgil spoke of the star and its light. Poets and poetess such as William Blake, Robert Frost, Letitia Elizabeth Landon, Alfred Lord Tennyson, and William Wordsworth wrote odes to it.

In Chinese, the planet is called Jin-Xing, the golden planet of the metal element. In India, Śukra Graha (planet Śukra) is named after the powerful saint Śukra. The word *śukra* is used in Indian Vedic astrology, which means "clear, pure" or "brightness, clearness" in Sanskrit. One of the *navagraha*s, it is held to affect wealth, pleasure and reproduction; it was the son of Bhr̥gu, preceptor of the *daitya*s, and *guru* of the *asura*s. The word *śukra* is also associated with semen or generation. Venus is known as Kejora in Indonesian and Malaysian Malay. Modern Chinese, Japanese, and Korean cultures refer to the planet literally as the "metal star", based on the five elements.

The Columbian Maya considered Venus to be the most important celestial body after the sun and the moon. They called it "Chac Ek or Noh Ek" (great star). The cycles of Venus were important to their calendar and were described in some of their books, such as *Maya Codex* of Mexico and *Dresden Codex*.

The ancient Egyptians and Greeks believed Venus to be two separate bodies, a morning star and an evening star. The Egyptians knew the morning star as Tioumoutiri and the evening star as Ouaiti. The Greeks used the names Phosphoros, meaning "light-bringer" (whence the element phosphorus); alternately Eosphoros, meaning "dawn-bringer", for the morning star, and Hesperos, meaning "western one", for the evening star. Though by the Roman era, they were recognized as one celestial object, known as "the star of Venus", the traditional two Greek names continued to be used, although usually translated into Latin as Lūcifer and Vesper.

Conclusion

Let us focus on the interesting facts about Venus:

1. A day on Venus is longer than a year. It takes Venus longer to rotate once on its axis than to complete one orbit of the sun. That is 243 earth days to rotate once — the longest rotation of any planet in the solar system — and only 224.7 earth days to complete an orbit of the sun.
2. Venus is hotter than Mercury — despite being further away from the sun. Its mean temperature is 462 °C. This is because of the high concentration of carbon dioxide in the atmosphere of the Venus, which works to produce an intense greenhouse effect. Heat is trapped in atmosphere like a blanket, causing the temperature of the planet to be much higher than its proximity to the sun would suggest.
3. Unlike the other planets in our solar system, Venus spins clockwise on its axis. All other planets spin anti-clockwise on their axis and orbit the sun in an anti-clockwise direction. Venus also orbits the sun anti-clockwise, but its unusual axis rotation is due to being upside down — it was knocked off its upright position earlier in its history!
4. Venus is the second brightest natural object in the night sky after the moon. The clouds of sulphuric acid in Venus'

atmosphere make it reflective and shiny, obscuring our view of its surface. Its brightness makes it visible even during the day – if it is clear and you know where to look.

5. Venus has 90 times the atmospheric pressure of the earth. That is about the same as the pressure found at a depth of 1 km in the earth's oceans.
6. Venus is named after the Roman Goddess of Love and Beauty. It is believed that Venus was named after the beautiful Roman Goddess (counterpart to the Greek Aphrodite) due to its bright, shining appearance in the sky. Of the first planets known to ancient astronomers, it would have been the brightest.
7. Venus was the first planet to have its motions plotted across the sky, as early as the second millennium BCE. Because Venus is easy to spot with the naked eye, it is impossible to say who discovered the planet. But over the centuries we have been able to measure Venus' motions, including the rare transit of Venus, when the planet appears to cross in front of the sun.
8. We tend to say, "Venusian", not "Venurian". Following the rules of Latin, we should say "Venurian" as the adjective to describe things related to Venus. However, this is deemed to be too close to the word "venereal". The more commonly used word is "Venusian" despite its chunky etymology.

35

Characteristic Features of the Moon

Introduction

THE moon is the only natural satellite of the earth. Together with the earth, it forms the earth–moon satellite system. It is about one-quarter of the earth in diameter (comparable to the width of Australia). In the solar system, it is the fifth largest satellite, larger than any of the known dwarf planets and the largest (and most massive) satellite of a planet relative to the planet itself. The moon is a planetary-mass object that formed a differentiated rocky body, making it a satellite planet under the geophysical definitions of the term. It lacks any significant atmosphere, hydrosphere, or magnetic field. Its surface gravity is about one-sixth of that of the earth (0.1,654 g). Jupiter's moon Lo is the only satellite in the solar system known to have a higher surface gravity and density.

Designation

Alternatively, the moon is known as "Luna, Selene (poetic), and Cynthia (poetic)". Likewise, the adjectives of the moon are lunar, selenian, cynthian and moonly.

Orbital Characteristics

Perigee	362,600 km (356,400-370,400 km)
Apogee	405,400 km (404,000-406,700 km)

Semi-major axis	384,399 km
Eccentricity	0.0549
Orbital period (sidereal)	27.321661 d
Orbital period (synodic)	29.5305589 d (29 d, 12 h, 44 min, 2.9 s)
Average orbital speed	1.022 km/s
Inclination	5.145° to the ecliptic
Longitude of ascending node	Regressing by one revolution in 18.61 years
Argument of perigee	Progressing by one revolution in 8.85 years

Satellite of the Earth

Physical Characteristics

Mean radius	1,737.4 km (0.2727 of earth's)
Equatorial radius	1,738.1 km (0.2725 of earth's)
Polar radius	1,736.0 km (0.2731 of earth's)
Flattening	0.0012
Circumference	10,921 km (equatorial)
Surface area	3.793×10^7 km^2 (0.074 of earth's)
Volume	2.1958×10^{10} km^3 (0.02 of earth's)
Mass	7.342×10^{22} kg (0.0123 of earth's)
Mean density	3.344 g/cm^3 (0.606 × earth)
Surface gravity	1.622 m/s^2 (0.1654 g)
Moment of inertia factor	0.3929 ± 0.0009
Escape velocity	2.38 km/s
Synodic rotation period	29.530589 d (29 d, 12 h, 44 min, 2.9 s; synodic; solar day)
Sidereal rotation period	27.321661 d
Equatorial rotation velocity	4.627 m/s

Axial tilt	1.5424° to ecliptic, 6.687° to orbit plane, 24° to earth's equator
North pole right	17 h, 47 m, 26 s
Ascension	266.86°
North Pole declination	65.64°
Albedo	0.136
Surface temperature	min mean max
at equator	100 °K 250 °K 390 °K
at 85° N	150 °K 230 °K
Surface absorbed dose rate	13.2 μGY/h
Surface equivalent dose rate	57.0 μSV/h
Apparent magnitude	(-2.5 to -12.9) (-12.74) (mean full moon)
Angular diameter	29.3 to 34.1 arc minutes

Atmosphere

Surface pressure	10^{-7} Pa (day), 10^{-10} Pa (night)
Composition by volume	He, Ar, Ne, Na, K, H, Rn

The moon orbit the earth at an average distance of 384,400 km or about thirty times the earth's diameter, its gravitational influence very slowly lengthens the earth's day and is the main driver of earth's tides. The moon's orbit around the earth has a sidereal period of 27.3 days. During each synodic period of 29.5 days, the amount of visible surface illuminated by the sun varies from zero to 100 per cent, resulting in lunar phases that form the basis for the months of a lunar calendar. The moon is tidally locked to the earth, which means that the length of a full rotation of the moon on its own axis causes its same side (the near side) to always face the earth, and the somewhat longer lunar day is the same on the synodic period. However, 59 per cent of the total lunar surface can be seen from the earth through shifts in perspective due to liberation.

The most widely accepted origin explanation posits that the moon formed 4.51 billion years ago, not long after earth, out of the debris from a giant impact between the planet and a hypothesized Mars-sized body called Theia. It then receded to a wider orbit because of tidal interaction with the earth. The near side of the moon is marked by dark volcanic maria (seas), which fill the space between bright ancient crustal highlands and prominent impact craters. Most of the large impact basins and mare surfaces were in place by the end of the Imbrian period, some 3 billion years ago. The lunar surface is relatively non-reflective, with a reflectance just slightly brighter than that of worn asphalt. However, because it has a large angular diameter, the full moon is the brightest celestial object in the night sky. The moon's apparent size is nearly the same as that of the sun, allowing it to cover the sun almost completely during a total solar eclipse.

Both the moon's prominence in the earth's sky and its regular cycle of phases have provided cultural references and influences for human societies throughout history. Such influences can be found in language, calendar systems, art, and mythology. The first artificial object to reach the moon was the Soviet Union's Luna-2, an uncrewed spacecraft in 1959; this was followed by the first successful soft landing by Luna-9 in 1966. The only human lunar missions to date have been those of the United States' Apollo programme, which landed twelve men on the surface between 1969 and 1972. These and later uncrewed missions brought lunar rocks that have been used to develop a detailed geological understanding of the moon's origins, internal structure, and subsequent history.

Physical Characteristics in Venus

The moon is a very slightly scalene ellipsoid due to tidal stretching, with its long axis displaced 30° from facing the earth, due to gravitational anomalies from impact basins. Its shape is more elongated than current tidal forces can account for. This "fossil bulge" indicates that the moon solidified when it orbited at half its current distance to the earth and that it is now too cold for its shape to adjust to its orbit.

Size and Mass

The moon is by size and mass the fifth largest natural satellite of the solar system, categorizable as one of its planetary-mass moons, making it a satellite planet under the geophysical definitions of the term. It is smaller than Mercury and considerably larger than the largest dwarf planet of the solar system, Pluto, though now Pluto is not considered as a planet. While the minor-planet moon Charon of the Pluto-charon system is larger relative to Pluto, the moon is the largest natural satellite of the solar system relative to their primary planets.

The moon's diameter is about 3,500 km, more than a quarter of the earth's diameter, with the face of the moon comparable to the width of Australia. The whole surface area of the moon is about 38 million sq. km, slightly less than the area of the Americas (North and South America put together).

The moon's mass is 1/81 of that of the earth, being the second densest among the planetary moons, and having the second highest surface gravity, after Lo, at 0.1654 g and an escape velocity of 2.38 km/s.

Internal Structure of the Moon

The moon is a differentiated body that was initially in hydrostatic equilibrium but has since departed from this condition. It has geochemically distinct crust, mantle, and core. The moon has a solid iron-rich inner core with a radius possibly as small as 240 km and a fluid outer core primarily made of liquid iron with a radius of roughly 300 km. Around the core is a partially molten boundary layer with a radius of about 500 km. This structure is thought to have developed through the fractional crystallization of a global magma ocean shortly after the moon's formation 4.5 billion years ago.

Crystallization of this magma ocean would have created a mafic mantle from the precipitation and sinking of the minerals olivine, clinopyroxene, and orthopyroxene; after about three quarters of the magma ocean had crystallized, lower-density plagioclase minerals could form and float into a crust atop. The final liquids to crystallize would have been initially sandwiched between the crust and the mantle, with a high abundance of incompatible and heat-producing elements. Consistent with this perspective, geochemical mapping made from the orbit suggests a crust of mostly anorthosite. The moon-rock samples of the flood lavas that erupted onto the surface from partial melting in the mantle confirm the mafic mantle composition, which is more iron-rich than that of the earth. The crust is on average about 50 km thick.

The moon is the second-densest satellite in the solar system, after Lo. However, the inner core of the moon is small, with a radius of about 350 km or less, around 20 per cent of the radius of the moon. Its composition is not well understood, but is probably metallic iron alloyed with a small amount of sulphur and nickel; analyses of the moon's time-variable rotation suggest that it is at least partly molten. The pressure at the lunar core is estimated to be 5 GPa (49,000 atm).

Magnetic Field of the Moon

The moon has an external magnetic field of generally less than 0.2 nanoteslas, or less than one hundred thousandth that of the earth. The moon does not currently have a global dipolar magnetic field and only has crustal magnetization likely acquired early in its history when a dynamo was still operating. However, early in its history, 4 billion years ago, its magnetic field strength was likely close to that of earth too. This early dynamo field apparently expired by about 1 billion years ago, after the lunar core had completely crystallized. Theoretically, some of the remnant magnetization may originate from transient magnetic fields generated during large impacts through the expansion of plasma clouds. These clouds are generated during large impacts in an ambient magnetic field. This is supported by the location of the largest crustal magnetizations situated near the antipodes of the giant impact basins.

Gravitational Field

The moon's gravitational field is not uniform, but on an average, at the moon's surface, it is 1.622 m/s^2 (0.1654 g) strong. Therefore, the surface gravity of the moon is about half of the surface gravity of Mars and about a sixth of the earth.

The details of the gravitational field have been measured through tracking the Doppler Shift of radio signals emitted by orbiting spacecraft. The main lunar gravity features are mescons, large positive gravitational anomalies associated with some of the giant impact basins, partly caused by the dense mare basaltic lava flows that fill those basins. The anomalies greatly influence the orbit of spacecraft about the moon. These are some puzzles: lava flows by themselves cannot explain all of the gravitational signature and some mascons exist that are not linked to mare volcanism.

Surface Conditions

The surface of the moon, having a surface pressure of 10^{-10} Pa, lacks an atmosphere which can regulate the resulting extreme conditions of the surface. The surface is exposed to drastic temperature differences ranging from 140 °C to -171 °C depending on the solar irradiance. Ionizing radiation from cosmic rays, the sun and the resulting neutron radiation produce radiation levels on average of 1,396 microsieverts per day, which is about 2-3 times more than on the International Space Station at about 400 km above earth in orbit, 5-10 times more than during a trans-Atlantic flight, 200 times more than on the earth's surface. For further comparison, radiation on a flight to Mars is about 1.84 millisieverts per day and on Mars 0.64 millisieverts per day. These extreme conditions, for example, are considered making it unlikely for spacecraft to harbour bacterial spores at the moon longer than just one lunar orbit. The surface gravity of the moon is approximately 1.625 m/s^2, about a sixth of 0.166 g of that on the earth's surface and about half on Mars's.

Surface Temperature

The moon's axial tilt with respect to the ecliptic is only 1.5427°, much less than that of the earth, i.e. 23.44°. Because of this small tilt, the moon's solar illumination varies much less with season than on the earth and it allows for the existence of some peaks of eternal light at the moon's north pole, at the rim of the crater peary.

Because of the lack of atmosphere, temperatures of different areas vary particularly upon whether they are in sunlight or shadow, making topographical details play a decisive role on local surface temperature. Parts of many craters, particulary the bottoms of many polar craters, are permanently shadowed; these "craters of eternal darkness" have extremely low temperatures. The Lunar Reconnaissance Orbiter measured the lowest summer temperatures in craters at the southern pole at 35 K (–238 °C) and just 26 K (–247 °C) close to the winter solstice in the north polar crater Hermite. This is the coldest temperature in the solar system ever measured by a spacecraft, colder even than the surface of Pluto.

Atmosphere of the Moon

The thin lunar atmosphere is visible on the moon's surface at sunrise and sunset with the lunar horizon glow and lunar twilight rays, like the earth's crepuscular rays. This Apollo-17 sketch depicts the glow and rays among the general zodiacal light.

The moon has an atmosphere so tenuous as to be nearly vacuum, with a total mass of less than 10 tonnes (9.8 long tons, 11 short tons). The surface pressure of this small mass is around 3×10^{-15} atm (0.3 nPa); it varies with the lunar day. Its sources include outgassing and sputtering, a product of the bombardment of lunar soil by solar wind ions. Elements that have been detected include sodium and potassium, produced by sputtering (also found in the atmospheres of Mercury and Lo; helium-4 and neon from the solar wind; and argon-40, randon-222, and polonium-210, outgassed after their creation by radioactive decay within the crust and mantle. The absence of such neutral species (atoms or molecules) as oxygen, nitrogen, carbon, hydrogen and magnesium, which are present in the regolith, is not understood. Water vapour has been detected by Chandrayaan-1 and found to vary with latitude, with a maximum at the order of 60-70 degrees; it is possibly generated from the sublimation of water ice in the regolith. These gases either return into the regolith because of the moon's gravity or are lost to space, either through solar radiation pressure or, if they are ionized, by being swept away by the solar wind's magnetic field.

Studies of the moon magma samples retrieved by the Apollo missions demonstrate that the moon had once possessed a relatively thick atmosphere for a period of 70 million years between 3 and 4 billion years ago. This atmosphere, sourced from gases ejected by lunar volcanic eruptions, was twice the thickness of that of present-day Mars. The ancient lunar atmosphere was eventually stripped away by solar winds and dissipated into space.

Dust Cloud

A permanent moon-dust cloud exists around the moon, generated by small particles from comets. Estimates are 5 tonnes of comet particles strike the moon's surface every 24 hours, resulting in the ejection of dust particles. The dust stays above the moon approximately 10 minutes, taking 5 minutes to rise and 5 minutes to fall. On average, 120 kg of dust are present above the moon, rising up to 100 km above the surface. Dust counts made by LADEE's Lunar Dust Experiment (LDEX) found particle counts peaked during the Geminid, Quadrantid, Northern Taurid, and Omicron Centaurid meteor showers, when the earth and the moon pass through the comet debris. The lunar-dust cloud is asymmetric, being more dense near the boundary between the moon's dayside and nightside.

Surface Features

The topography of the moon has been measured with laser altimetry and stereo image analysis. Its most extensive topographic feature is the giant far-side South Pole–Aitken Basin, some 2,240 km in diameter, the largest crater on the moon and the second-largest confirmed impact crater in the solar system. At 13 km deep, its floor is the lowest point on the surface of the moon. The highest elevations of the moon's surface are located directly to the north-east, which might have been thickened by the oblique formation impact of South Pole–Aitken Basin. Other large impact basins such as Imbrium, Serenitatis, Crisium, Smythii, and Orientale possess regionally low elevations and elevated rims. The far-side of the lunar surface is on average about 1.9 km higher than that of the near side.

The discovery of fault scarp cliffs suggests that the moon has shrunk by about 90 m within the billion years. Similar shrinkage features exist on Mercury. Mare Frigoris, a basin near the north pole long assumed to be geologically dead, has cracked and shifted. Since the moon does not have tectonic plates, its tectonic activity is slow and cracks develop as it loses heat.

Volcanic Features

The main features visible from the earth by the naked eye are dark and relatively featureless lunar plains called maria (singular mare; Latin for "seas", as they were once believed to be filled with water) are vast solidified pools of ancient basaltic lava. Although similar to terrestrial basalts, lunar basalts have more iron and no minerals altered by water. The majority of these lava deposits erupted or flowed into the depressions associated with the impact base. Several geologic provinces containing shield volcanoes and volcanic domes are found within the near side "maria".

Almost all maria are on the near side of the moon and cover 31 per cent of the surface of the near side compared with 2 per cent of the far side. This is likely due to a concentration of heat-producing elements under the crust on the near side, which would have caused the underlying mantle to heat up, partially melt, rise to the surface and erupt. Most of the moon's mare basalts erupted during the Imbrian period, 3.3-3.7 billion years ago, though some being as young as 1.2 billion years and as old as 4.2 billion years.

In 2006, a study of Ina, a tiny depression in Lacus Felicitatis, found jagged, relatively dust-free features that, because of the lack of erosion by infalling debris, appeared to be only 2 million years old. Moonquakes and releases of gas indicate continued lunar activity. Evidence of recent lunar volcanism has been identified at 70 irregular mare patches, some less than 50 million years old. This raises the possibility of a much warmer luanr mantle than previously believed, at least on the near side where the deep crust is substantially warmer because of the greater concentration of radioactive elements. Evidence has been found for 2-10 million year old basaltic volcanism within the crater Lowell, inside the Orientale Basin. Some combination of an initially hotter mantle and local enrichment of heat-producing elements in the mantle could be responsible for prolonged activities on the far side in the Orientale Basin.

The lighter-coloured regions of the moon are called terrae, or more commonly highlands, because they are higher than most maria. They have been radiometrically dated to having formed 4.4 billion years ago, and may represent plagioclase cumulates of the lunar magma ocean. In contrast to earth, no major lunar mountains are believed to have formed as a result of tectonic events.

The concentration of maria on the near side likely reflects the substantially thicker crust of the highlands of the far side, which may have formed in a slow-velocity impact of a second moon of the earth a few tens of millions of years after the moon's formation. Alternatively, it may be a consequence of asymmetrical tidal heating when the moon was much closer to the earth.

Impact Craters

A major geologic process that has affected the moon's surface is impact cratering, with crates formed when asteroids and comets collide with the lunar surface. There are estimated to be roughly 300,000 craters wider than 1 km on the moon's near side. The lunar geologic timescale is based on the most prominent impact events, including Nectaris, Imbrium, and Orientale structures characterized by multiple rings of uplifted material between hundreds and thousands of kilometres in diameter and associatd with a broad apron of ejecta deposits that form a regional stratigraphic horizon. The lack of an atmosphere, weather, and recent geological processes mean that many of these craters are well-preserved. Although only a few multi-ring basins have been definitively dated, they are useful for assigning relative ages. Because impact craters accumulate at a nearly constant rate, counting the number of craters per unit area can be used to estimate the age of the surface. The radiometric ages of impact-melted rocks collected during the Apollo missions cluster between 3.8 and 4.1 billion years old; this has been used to propose a Late Heavy Bombardment period of increased impacts.

High-resolution images from the Lunar Reconnaissance Orbiter in the 2010s show a contemporary crater production rate significantly higher than what the previous estimation. A secondary cratering process caused by distal ejecta is thought to churn the top 2 cm of regolith on a timescale of 81,000 years. This rate is 100 times faster than the rate computed from models based solely on direct micrometeorite impacts.

Lunar Swirls

Lunar swirls are enigmatic features found across the moon's surface. They are characterized by a high albedo, appear optically immature (i.e. the optical characteristics of a relatively young regolith and often have a sinuous shape. Their shape is often accentuated by low albedo regions that wind between the bright swirls. They are located in places with enhanced surface magnetic fields and many are located at the antipodal point of major impacts. Well-known swirls include the Reiner Gamma feature and Mare Ingenii. They are hypothesized to be areas that have been partially shielded from the solar wind, resulting in slower space weathering.

Surface Composition

Blanketed on top of the moon's crust is a highly comminuted (broken into ever smaller particles) and impact gardened mostly gray surface layer called regolith, formed by impact processes. The finer regolith, the lunar soil of silicon dioxide glass, has a texture resembling snow and a scent-resembling spent gunpowder. The regolith of older surfaces is generally thicker than that of younger surfaces; it varies in thickness from 10-15 m in the highlands and 4-5 m in the maria. Beneath the finely comminuted regolith layer is the megaregolith, a layer of highly fractured bedrock many kilometres thick.

Retalive molecular composition of the lunar surface is given below:

Compound	*Formula*	*Composition*	
		Maria (%)	*Highlands (%)*
Silica	SiO_2	45.4	45.5
Alumina	Al_2O_3	14.9	24.0
Lime	CaO	11.8	15.9
Iron (ii) oxide	FeO	14.1	5.9
Magnesia	MgO	9.2	7.5
Titanium dioxide	TiO_2	3.9	0.6
Sodium oxide	Na_2O	0.6	0.6
Total		99.9	100.0

Presence of Water

Liquid water cannot persist on the lunar surface. When exposed to solar radiation, water quickly decomposes through a process known as 64 and is lost to space. However, since the 1960s, scientists have hypothesized that water-ice may be deposited by impacting comets or possibly produced by the reaction of oxygen-rich lunar rocks and hydrogen from solar wind, leaving traces of water which could possibly persist in cold, permanently shadowed craters at either pole on the moon. Computer simulations suggest that up to 14,000 km^2 of the surface may be in permanent shadow. The presence of usable quantities of water on the moon is an important factor in rendering lunar habitation as a cost-effective plan; the alternative of transporting water from earth would be prohibitively expensive. In course of studying the samples of dust collected from the moon, signs of water have been found to exist on the lunar surface. In 1994, the bistatic radar experiment located on the clementine spacecraft, indicated the existence of small, frozen pockets of water close to the surface. However, later radar observations by Arecibo suggest that these findings may rather be rocks ejected from young impact craters. In 1998, the neutron spectrometer on the lunar prospector spacecraft showed that high concentrations of hydrogen are present in the first metre of depth in the regolith near the polar regions volcanic lava beads, brought back to the earth aboard Apollo-15, showed small amounts of water in their interior.

The 2008 Chandrayaan-1 spacecraft has since confirmed the existence of surface water-ice, using the on-board Moon Mineralogy Mapper. The spectrometer observed absorption lines, common to hydroxyl, in reflected sunlight, providing evidence of large quantities of water ice, on the lunar surface. The spacecraft showed that concentration may possibly be as high as 1,000 ppm. Using the mapper's reflectance spectra, indirect lighting of areas in shadow confirmed water-ice within 20° latitude of both poles in 2018. In 2009, LCROSS sent a 2,300 kg impactor into a permanently shadowed polar crater and detected at least 100 kg of water in a plume of ejected material. Another examination of the LCROSS data showed the amount of detected water to be closer to 155 ± 12 kg.

In May 2011, 615-1410 ppm water in melt inclusions in lunar

sample 74220 was reported, the famous high-titanium "orange glass soil" of volcanic origin collected during the Apollo-17 mission in 1972. The inclusions were formed during explosive eruptions on the moon approximately 3.7 billion years ago. This concentration is comparable with that of magma in earth's upper mantle. Although of considerable selenological interest, this insight does not mean that water is easily available since the sample was many kilometres below the surface and the inclusions are so difficult to access that it took thirty-nine years to find them with a state-of-the-art ion microprobe instrument.

Analysis of the findings of the Moon Mineralogy Mapper (M3) revealed in August 2018 for the first time "definitive evidence" of water-ice on the lunar surface. The data revealed the distinct reflective signatures of water-ice, as opposed to dust and other reflective substances. The ice deposits were found on the North and South poles, although it is more abundant in the south, where water is trapped in permanently shadowed craters and crevices, allowing it to persist as ice on the surface since they are shielded from the sun.

In October 2020, astronomers reported detecting molecular water on the sunlit surface of the moon by several independent spacecraft, including the Stratospheric Observatory for Infrared Astronomy (SOFIA).

The Moon in Culture and Life

Here, Lunar Calendar, Lunisolar Calendar, Metonic Cycle, Blue Moon, Black Moon, and Movable Feast are discussed.

CALENDAR

Since pre-historic times people have taken note of the moon's phases, its waxing and waning, and used it to keep record of time. Tally sticks, notched bones dating as far back as 20,000-30,000 years ago, are believed by some to mark the phases of the moon. The counting of the days between the moon's phases gave eventually rise to generalized time periods of the full lunar cycle as months and, possibly, of its phases as weeks.

The words for the month in a range of different languages carry this relation between the period of the month and the moon etymologically. The English month as well as the moon and its cognates in other Indo-European languages (e.g. the Latin *mensis* and ancient Greek *meis* or *mēn*, meaning "month" stem from the Proto-Indo-European (PIE) root of the moon. Similarly, the Chinese language uses the same word for moon as well as for month, which furthermore can be found in the symbols for the word week.

This lunar timekeeping gave rise to the historically dominant, but varied, lunisolar calendars. The seventh-century Islamic calendar is an example of a purely lunar calendar, where months are traditionally determined by the visual sighting of the hilal, or earliest crescent moon over the horizon. There is particular significance for a range of calendars and cultures, on the occasion of the full moon, to use or celebrate, particularly around the autumnal equinox, the so-called Harvest Moon. Moreover, association of time with the moon can also be found in religions such as the ancient Egyptian temporal and lunar deity Khonsu. Ostensibly, the moon is worshipped as a deity (Goddess) in the various cultures of the world. In modern times, the moon has been represented in science fictions also.

Conclusion

The perception of the moon in modern times has been informed by telescope-enabled modern astronomy and later by spaceflight-enabled actual human activities, at the moon, particularly the culturally impactful lunar landings. These new insights inspired cultural references, connecting romantic reflections about the moon and speculative fictions such as science fiction dealing with the moon.

Contemporarily the moon has been seen as a place for economic expansion into space, with missions prospecting for lunar resources. This has been accompanied with renewed public and critical reflection on humanity's cultural and legal relation to the celestial body, especially regarding colonialism, as in the 1970 poem "Whitey on the Moon". In this light, the moon's nature has been invoked, particularly for lunar conservation. Various authors and poets of the world have long written about the beauty of the moon, in particular, its beams.

36

Epilogue

Most of the space activities have been pursued since they serve some utilitarian purpose, whether increasing knowledge, adding to national power, or making some profit. Nevertheless, there remains a powerful underlying sense that it is important for humans to explore space for their own take, "to see what is there". Although the only voyages that humans have made away from the near vicinity of the earth — the Apollo flights to the moon — were motivated by Cold War competition, there have been recurrent calls for humans to go to the moon, travel to Mars, and visit other locations in the solar system and beyond. Until humans resume such journeys of exploration, robotic spacecraft will continue to serve in their stead to explore the solar system and probe the mysteries of the universe.

The first artificial earth satellite, Sputnik-1, was launched by the Soviet Union on 4 October 1957. The first human to go into space, Yuri Gagarin, was sent again by the Soviet Union for a one-orbit journey around the earth on 12 April 1961. Within ten years of that first human flight, American astronauts walked on the surface of the moon. Apollo-11 crew members Neil Armstrong and Edwin (Buzz) Aldrin made the first lunar landing on 20 July 1969. A total of twelve Americans on six separate Apollo missions set foot on the moon between July 1969 and December 1972. Since then, no humans have left earth orbit, but more than 500 men and women have spent as many as 438 consecutive days in space. Starting in the early 1970s, a series of Soviet (Russia from December 1991) space stations, the US Skylab Station, and numerous space shuttle flights provided earth-orbiting bases for varying periods of human occupancy and activity. From 2 November 2000, when its first crew took up residence, to its completion in 2011, the International Space Station (ISS) served as a base for humans living and working in space on a permanent basis. It will continue to be used in this way at least until 2024.

Since 1957, earth-orbiting satellites and robotic spacecraft, journeying away from the earth, have gathered valuable data about the sun, the earth, other bodies in the solar system, and the universe beyond. Robotic spacecraft have landed on the moon, Venus, Mars, Titan, a comet, and four asteroids have visited all the major planets, and have flown by Kuiper belt objects and by the nuclei of comet including Halley's Comet, travelling the inner solar system. Scientists have used space-derived data to deepen human understanding of the origin and evolution of galaxies, stars, planets, and other cosmological phenomena.

Orbiting satellites also have provided, and continue to provide, important services to the everyday life of many people on the earth. Meteorologic satellites deliver information on short-and long-term weather patterns and their underlying causes. Other earth-observation satellites remotely sense land and ocean areas, gathering data that improve management of earth's resources and that help in understanding global climate change. Telecommunications satellites allow essentially instantaneous transfer of voice, images, and data on a global basis. Satellites operated by Space Agencies of the United States, Russia, China, Japan, India, and Europe give precision navigation, positioning, and timing information that has become essential to many terrestrial users. Earth-observation satellites have also become extremely useful to the military authorities of several countries as complements to their army, navy and air forces and have provided important security-related information to national leaders.

As the many benefits of space activity became evident, other countries have joined the Soviet Union (Russia) and the United States in developing their own space programmes. They include a number of Western European countries operating both individually and, after 1975, cooperatively through the European Space Agency, as well as China, Japan, Canada, India, Israel, Iran, North Korea, South Korea, and Brazil. By the second decade of the twenty-first century, more than fifty countries had space agencies or other government bodies carrying out space activities.

Bibliography

Basu, Biman, 2006, *Cosmic Vistas*, repr. edn, New Delhi: National Book Trust.

Biswal, S. and B.L. Ray, 2009, *Vedic Science and Technology*, New Delhi: D.K. Printworld.

Biswal, S. and P. Paikray, 2017, "Scramjet Technology in Indian Rocketery", *Emerging Science* 9 (August): 21-23.

———, 2018, "Way of Detecting a New Planet in the Celestial Space", *Emerging Science* 10 (February): 23-24.

———, 2018a, "Women in Space Science", *Emerging Science* 10 (October-December): 27-28.

Biswal, S., 2000, *Understanding Some Aspects of Physics*, Delhi: Emkay Publications.

Biswal, S., 2005, *Theory of Relativity*, Delhi: Vrinda Publications.

———, 2016, "Aditya-1: India's Maiden Mission to the Sun", *Emerging Science*, 8(4) (December): 7-9.

———, 2016a, "Discovery of Gravitational Waves", *Science Horizon*, 6(3) (March): 3-6.

———, 2016b, "Sustenance of Life with Light", *Emerging Science*, 8(1) (March): 28-30.

———, 2016c, *Exploring Physics Behind*, Bhubaneswar: Bhadra Prakashani.

———, 2017, "Detection of Evidence for the Existence of New Planets", *Science Horizon*, 2(10) (October): 45-47.

———, 2017a, "Outlines of Global Positioning System", *Science Horizon*, 2(9) (September): 11-12.

———, 2017b, "Solar Energy: Genesis, Growth and Useful Technical Methods", National Seminar on Solar Energy, held on 26-27 March 2017, Organized by the Department of Chemistry, P.N. College, Bolgarh, Khodha, Odisha.

———, 2018, "Physics as Conceptualized in Ancient India" *Science Horizon*, 3(12) (December): 27-33.

———, 2021, "Einstein's Universe", *Emerging Science*, 13(1) (January-March): 9-10.

———, 2021a, "NASA's Five Significant Missions", *Current Development Reporter*, VII(4) (October-December): 11-16.

———, 2021b, "Physics of the Earth's Atmosphere", *Emerging Science* 13(4) (October-December), 10: 8-10.

———, 2021c, "Role of Physics in Information and Communication Technology", *Current Development Reporter*, VII(2-3) (April-September): 21-24.

———, 2021d, "Satellites: An Overview", *Science Horizon* 6(11) (November): 516-19.

———, 2022, "Moon Mission: A Bird's-Eye View", *Current Development Reporter*, VIII(1) (January-March): 19-25.

Biswal, S., M. Jena, and B. Pradhan, 2020, *Amazing Facts of Physics*, Bhubaneswar: Lakhmi Prakashans.

Capra, Fritjof, 1982, *The Tao of Physics*, 3rd edn, Flamingo, FL, Harper Collins Publishers.

Dongre, N.G., 1994, *Physics in Ancient India*, New Delhi: New Age International Publishers.

Chandra Bhushan, 2016, "How the World become Flat" *Down to Earth*, annual number CSE, New Delhi, 16-29 February, pp. 12-16.

Everyman's *Science*, 16 October–16 November 2016, LI(4): 72-73.

Kapila, R. and A. Kapila, 1992, *Concise Encyclopaedia of Nobel Laureates*, Delhi: Academic Foundation.

Lange, F.A., 1957, *The History of Materialism*, tr. E.C. Thomas, London: Routledge and Kegan Paul.

Landau, L. and Yu. Rumer, 1961, *What Is the Theory of Relativity*, Moscow: Peace Publishers.

Muralidhar, K. and T. Sundararajan, 1995, *Computational Fluid Flow and Heat Transfer*, New Delhi: Narosa Publishing House.

Murthy, S.R.N., 1997, *Vedic View of the Earth*, New Delhi, D.K. Printworld.

Murthi, R.K., 2005, *The Great Aviation Story*, 1st edn, New Delhi: National Book Trust.

Oxford Dictionary of Science, 6th edn, Oxford University Press, 2010.

Panda, N.C., 2008, *Maya in Physics*, repr edn, Delhi: Motilal Banarsidass.

Rajan, M.S., 2013, *Space Today*, repr. edn, New Delhi: National Book Trust.

Rajan, Mohan Sundara, 2013, *Space Today*, New Delhi: National Book Trust, repr edn, pp. 215-24.

Rajaram, N.S, 1993, "Vedic and Harappan Culture: New Findings", *Jour. Mythic Soc.*, 84: 133-37.

Ray Chowdhury, S.C. and D.B. Sinha, 1946, *Intermediate Physics*, vol. 1, 17th repr. edn, Kolkata: Modern Book Agency.

Ronan, C.A., 1964, *The Astronomers*, London: Evans Brothers.

Russell, B., 1954 (2021), *A History of Western Philosophy*, London: George Allen and Unwin.

Singh, Bal Ram et al. (eds), 2012, *Science and Technology in Ancient Indian Texts*, New Delhi: D.K. Printworld.

Subbarayappa, B.V. and K.V. Sarma, 1985, *Indian Astronomy: A Source Book*, Bombay: Nehru Centre.

Thibaut, G. and S. Dwivedi, 1968, *Pañcasiddhāntikā of Varāhamihira*, Varanasi: Chowkhamba Sanskrit Series.

Venkateswaran, T.V., 2016, "ISRO Conducts Successful In-fligt Test of Scramjet Engines", *Science Reporter*, October. pp. 54-55.

Weber, R.L., K.V. Manning, M.W. White and G.A. Weygand, *College Physics*, Part 1, Delhi: Tata McGraw-Hill.

Websites

http://www.esa.int>esapub,pdf, Down to Earth – European Space Agency.

http://www.sciencedaily.com/releases/2015

https//en.m.wikipedia.org – List of Missions to the Moon

https//en.m.wikipedia.org.wiki – Moon Landing – Wikipedia

https://en.m.wikipedia.org

https://en.m.wikipedia.org – List of Missions to the Moon

https://en.m.wikipedia.org > wiki–Moon – Wikipedia

https://en.m.wikipedia.org >wiki–Venus – Wikipedia

https://en.m.wikipedia.org.wiki – Moon Landing-Wikipedia

https://northeasternchronicle.in > news – Women in Space: Meet Indian Women Astronauts who Made History.

https://nssdc.gsfc.nasa.gov.planetary

https://solarsystem.nasa gov.>planets – Overview/Venus – NASA Solar System Exploration

https://www.britanica.com > place – Moon-Distinctive features, Britanica

https://www.esa.int, >esapub–PDF, *Down to Earth* – European Space Agency.

https://www.esa.int.bullet103:PDF – Ways to the Moon–European Space Agency

https://www.nasa.gov.missions

https://www.researchgate.net, 22119

https://www.rmg.co.uk – Interesting facts about Venus

https://www/esa.int, RACCA, PDF, The Smart-1 mission – European Space Agency.

Index

DK PRINTWORLD